Lecture Notes in Mathematics   1644

Editors:
A. Dold, Heidelberg
F. Takens, Groningen

Springer
Berlin
Heidelberg
New York
Barcelona
Budapest
Hong Kong
London
Milan
Paris
Santa Clara
Singapore
Tokyo

Allan Adler   S. Ramanan

Moduli of
Abelian Varieties

 Springer

Authors

Allan Adler
Cherokee Station
P.O. Box 20276
New York, NY 10021, USA
e-mail: adler@pulsar.cs.wku.edu

Sundararaman Ramanan
Tata Institute of Fundamental Research
Homi Bhabha Road
400 005 Mumbai, India
e-mail: ramanan@tifrvax.tifr.res.in

Cataloging-in-Publication Data applied for

Die Deutsche Bibliothek - CIP-Einheitsaufnahme

Adler, Allan:
Moduli of Abelian varieties / Allan Adler ; S. Ramanan. -
Berlin ; Heidelberg ; New York ; Barcelona ; Budapest ; Hong
Kong ; London ; Milan ; Paris ; Santa Clara ; Singapore ;
Tokyo : Springer, 1996
  (Lecture notes in mathematics ; 1644)
  ISBN 3-540-62023-0
NE: Ramanan, Sundararaman:; GT

Mathematics Subject Classification (1991): 11D41, 11F27, 11F32, 11F46, 11G05,
11G18, 14-02, 14D20, 14E15, 14J30, 14J35, 14J60, 14K10, 14K25, 14M12,
14M15, 14N05, 14N10, 16W99, 20C33, 20C34

ISSN 0075-8434
ISBN 3-540-62023-0 Springer-Verlag Berlin Heidelberg New York

Typesetting: Camera-ready T$_E$X output by the first author
SPIN: 10520167   46/3142-543210 - Printed on acid-free paper

Table of Contents

What this book is about

Let X be an abelian variety and let L be a very ample line bundle on X. Igusa [Ig] and Mumford [Mu1,I-III], [Mu2-4] have studied the embedding of X in the projective space associated to L. By considering the image of the identity element of X under the embedding, they also obtained the moduli of abelian varieties with a totally symmetric line bundle and a theta structure for that line bundle. Their results on the equations defining abelian varieties depend on the assumption that the line bundle is the fourth power of another line bundle while for the equations defining the moduli scheme they require the line bundle to be an eighth power. In particular, their results do not apply to the elliptic normal curves of odd degree and the moduli space of elliptic curves of odd level.

It is not the case that we have succeeded where they have failed. What we have done is to generalize some of the constructions of [Mu1,I] to arrive at equations similar to those which Mumford finds but for other levels, notably odd levels. This is fairly routine and doesn't involve any new ideas apart from the notion of a strongly symmetric line bundle (Def. 11.5). In no case have we proved that the equations we obtain define the moduli space. Instead, it is more accurate to say that we adapt Mumford's techniques to obtain equations defining loci whcih we then study for their own sake. In most cases, the equations we obtain were already known in the last century and were due to Felix Klein or and to Heinrich Burckhardt as relations among theta functions. It is nevertheless satisfying that we are able to get Mumford's machinery to yield these classical equations.

What is probably most interesting and original in this book is the attention we give to the loci that arise and the techniques which we have developed for studying them. The ideal situation would be the one in which we were able to prove general results about moduli spaces given only their equations in a general form and without using the fact that they are moduli spaces. To some extent we are successful with this. For example, using only Klein's equations for his z-curve of level p, we can compute the degree of the curve in general just from the equations. A refinement of these methods, again using just the equations, shows that if Fermat's last theorem is true for the prime $p \geq 7$, then the modular curve of level p has exactly $(p-1)/2$ rational points. But despite these successes, we cannot even compute the genus of the curve without knowing that it arises from the moduli space and that the moduli space has a certain genus. Similarly, we cannot prove that the locus is smooth in general or even that it does not have isolated points, whereas Vélu [Vé1] has been able to do this using the fact that he is working with moduli spaces.

One of the other successes of our approach is that we are able to give novel interpretations of classical equations defining moduli spaces and to generalize them in some cases. For example, we can interpret Klein's equations for the modular curve of level p (Theorem 19.17) as saying that the modular curve lies in the intersection of a Grassmannian and a 2-uply embedded projective space. In the special case $p = 7$, this amounts to the well-known observation that a plane quartic curve arises from the intersection of a Veronese surface with a quadric in $\mathbf{P}^5$. That Klein's equations actually define the moduli space $X(p)$ in characteristic different from p was proved by Vélu [Vé1], so the modular curve is precisely the intersection of a Grassmannian and a Veronese variety. Similarly, we can interpret theta relations of genus g and level 3 with trivial Arf invariant as saying that the moduli space maps to the rank $\leq 2^g$ locus of the Hessian of the unique quartic invariant for $Sp_{2g}(\mathbf{F}_3)$ acting on the odd part of the Weil representation.

One of the features of our approach is a greater use of the Weil representation [W1]. Indeed, the key to both the equations of Klein [K1-4], [K-F] and those of Burckhardt [Bu] lies in the use of a certain intertwining operator on the 2nd tensor power of the Weil representation. Accordingly we call that operator the **fundamental intertwining operator** (cf. (20.22) and (21.31)). However, we didn't actually realize the importance of this operator in its own right when we collaborated on this work, nor were we careful about distinguishing between the Weil representation *qua* projective representation and the lifting of the Weil representation to a linear representation. This defect is to some extent remedied by the inclusion of one of the first author's unpublished article [A1] as an appendix. The fundamental intertwining operator is discussed further in an appendix to the article [A6].

One of the advantages of having explicit equations for the moduli spaces is that it makes it possible to generalize theorems about moduli spaces to theorems about other classes of varieties of which the moduli spaces are special cases. This has been one of the concerns of much of the first author's research, including in the seminal article [A2] which led to this collaboration. The discussion in §22 alludes to the work of the first author in this direction. Furthermore, among the appendices will be found two preprints [A15], [A16] of the first author which illustrate the kind of generalization of which mention has just been made. The reader is also referred to the article [A6].

There are striking examples in the literature of geometrical treatments of equations defining moduli spaces, for example the book of Baker [Ba] on the moduli space of abelian surfaces of level $(3, 3)$ and its realization as a quartic threefold in four dimensional space and the papers [E1], [E2] of W.L.Edge. Because one can use the equations to good advantage in special cases and in general arguments, it follows that one must sharpen the available tools for working with the equations and the relevant geometry. One very promising tool is the study of rings of invariants, as Klein already noted. One example of the application of this tool is the first author's unpublished article [A11], which is included as an appendix. Another is the assistance offered by powerful computers and software, including the programs REDUCE, MACSYMA, CAYLEY and MACAULAY. While such software tools are better than nothing, and even quite good for many general purposes, they are still unfortunately in too primitive a state to be of proper assistance in our investigations. Hopefully, the situation will soon be improved.

Since Klein's projective embeddings of the modular curve, known as the z-curve and the A-curve, are in general not multicanonical embeddings, as one sees from their degrees, it is natural to ask what are the line bundles on $X(p)$ associated to these embeddings. This raises the general question of determining the invariant line bundles on the modular curve and this is carried out in §24. We show that the group of invariant line bundles is an infinite cyclic group generated by an invariant line bundle of degree $(p^2-1)/24$. Furthermore, our interpretation of Klein's equations for the z-curve leads us naturally to an invariant rank 2 vector bundle on $X(p)$, so we also discuss invariant vector bundles. The first author has also found geometrical ways to construct other vector bundles on certain modular curves (cf. Appendices IV and V).

Structure of the book

The structure of the book is as follows. In the first chapter, we discuss the properties of standard Heisenberg groups, of their representations and of certain homomorphisms between them. In Chapter II, we present a discussion of the Heisenberg groups of ample line bundles on abelian varieties which is essentially parallel to the discussion of Chapter I.

In Chapter III, we draw a transversal between the two parallel discussions by introducing the notion of a strongly symmetric theta structure. Using this notion, we are able to obtain in §17 the addition formula (17.17) and the duplication formula (17.30) and in §18 the inversion formula (18.25), the product formula (18.16) and the fundamental relation among the theta constants (18.7). In doing so, we follow the ideas of Mumford [Mu1,I] while taking into account the complications deriving from the presence of a nontrivial quadratic form on the half period group of the abelian variety.

In Chapter IV, we study the fundamental relations in the case of line bundles of odd level. We pay particular attention to the case of the modular curve of prime level and to the moduli of abelian varieties with a line bundle of level $\delta = (3, 3 \ldots, 3)$. In the latter case, we generalize Burkhardt's quartic invariant to a quartic invariant of the symplectic group $Sp(\mathbf{F}_3^{2g})$, which we also write down explicitly. The existence of an invariant tensor invariant of degree 4 was known previously to Roger Howe, who mentioned it in his fundamental unpublished paper [Ho] on dual reductive pairs but who considered neither the symmetry of the tensor nor its explicit form. This result was obtained by the authors during discussions at UCLA in the winter of 1979. It was obtained independently by Van der Geer [Vdg2]. The determination of the precise automorphism groups of these quartics was carried out in [A17].

Chapter V deals with miscellaneous relations of our work to invariant theory, number theory and to automorphic forms. In §22, we draw attention to the role of invariants of finite groups in these investigations and to some of the philosophical issues they present in connection with the relevant geometry. These issues were raised in the article [A2] and are discussed at greater length, with appropriate mathematics, in the article [A6] and in the lecture notes [A18]. In §23 we show, using only the equations of the modular curve of prime level, that if the modular curve of level $p \geq 7$ has a nontrivial rational point, then so does the Fermat curve of degree p. This result was obtained by the authors in discussions at the Institute for Advanced Study in 1977-78 and was directly inspired by Hurwitz's classical proof of the corresponding result for the case of the Klein curve

$$X^3Y + Y^3Z + Z^3X = 0.$$

The same result was obtained, and in stronger form, independently by Vélu [Vé2] and by Kubert [Ku]. However, our proof uses only a direct and naive analysis of the equations themselves, along Hurwitz's lines, and uses no information from the theory of elliptic curves. Again, it shows that one can work with the equations themselves. At the time we did this, Fermat's Last Theorem was still a conjecture. It has since been proved by Wiles [Wi1].

In §24, we study the vector bundles on the modular curve of prime level which are invariant under the group $SL_2(\mathbf{F}_p)$. As part of that study, we classify the line bundles on the modular curve of prime level which are invariant under the group $SL_2(\mathbf{F}_p)$. We show that the isomorphism classes of invariant line bundles form an infinite cyclic group generated by an invariant line bundle λ of degree $(p^2 - 1)/24$. In particular, the canonical line bundle on the modular curve is the N-th power of λ, where $N = 2p - 12$ and accordingly the canonical bundle has a canonical N-th root. For example, for $p = 11$, the canonical line bundle on the modular curve has a canonical 10th root. It therefore seems appropriate, as noted in [A2] and [A14], to regard the sections of powers of λ as modular forms of fractional weight. In [A14], some modular forms of weight 4/5 for $\Gamma(11)$ that arise in this way were computed explicitly. The first author's unpublished article [A14] is included here as an appendix.

By way of contrast, we know much less about invariant vector bundles of higher rank on modular curves, particularly the semistable vector bundles. Nevertheless we are able to comment on the rank 2 vector bundles on $X(p)$ which arise naturally from our study of Klein's equations for the z-curve.

The five chapters are supplemented by five appendices which are reproductions of five unpublished articles [A3], [A14], [A11], [A15], [A16] of the first author, slightly modified to achieve consistency with the pagination, section numbering and bibliography of the book as a whole. Also, material in the articles which duplicates material in the rest of this book was eliminated.

One regrettable lacuna in this book is that we do not address the question of the arithmetic groups which belong to the moduli spaces we consider.

For the convenience of the reader, we have also included an index of notation and a subject index.

Some Notational Conventions and Terminology

Throughout this volume, k will denote an algebraically closed field. Furthermore, we will assume that the characteristic of the field k is $\neq 2$, unless the contrary is explicitly indicated. When we deal with the case of level $(d_1, d_2, \ldots, d_r)$, we will assume that the characteristic of the field k does not divide any of the integers d_i.

In order to define our context somewhat, let us specify that throughout the book, an **algebraic variety** will mean a reduced integral scheme of finite type over the algebraically closed field k. By an **abelian variety** we will mean a group object in the category of complete varieties over k. By an **element** of an algebraic variety we will mean a closed point of that variety. Thus the elements of an abelian variety form a group. We will speak of the **kernel** of a homomorphism between two abelian varieties on various occasions. In view of our separability assumptions and other assumptions on the characteristic, the kernels in question will always arise from finite discrete groups in the set theoretic sense. Although we allude to some descent properties of line bundles with respect to certain kernels in §9, for the most part, we will primarily be concerned with these kernels as finite subgroups of the group of elements of an abelian variety and not as subvarieties or subschemes or sub group schemes of the abelian variety.

Additional comments and acknowledgments of the first author

The first appendix is my article [A3], which describes explicitly the lifting of Weil's projective representation of the finite symplectic group $Sp(2n, q)$ to a linear representation. In view of the importance of the Weil representation for this work, it seemed appropriate to include this. In [A3], Weil's original notation and terminology are retained. This approach is carried further in an appendix to [A6]. The reader interested in a different and more general presentation of the result of [A3] is referred to the article of Gérardin [Gé].

Appendix II is the article [A14], in which explicit functions on the upper half plane are computed which correspond to sections of the unique $PSL_2(\mathbf{F}_{11})$ invariant line bundle of degree 20 on $X(11)$. Since an invariant line bundle on $X(11)$ is uniquely determined by its degree, it is justifiable on geometric grounds to describe these functions as "automorphic forms of weight 4/5 for $\Gamma(11)$". However, it must be emphasized that these automorphic forms do not lend themselves to a treatment based on adelic metaplectic groups and reciprocity laws. The technique for writing down these automorphic forms is based on the bicycle structure of the ring of invariants computed in [A9], on character computations using the Woods Hole fixed point formula and on explicit theta expansions used by Klein to realize $X(11)$ as curves of degrees 20 and 50 in $\mathbf{P}^4$.

Appendix III is the little paper [A11] which serves to illustrate some of my ideas about the use of invariant theory in the study of moduli spaces and other varieties with large automorphism groups. In it, the results of my papers [A8],[A9], where the ring of invariants of a 5 dimensional complex representation of $PSL_2(\mathbf{F}_{11})$ was computed, are applied to determine equations that define set-theoretically Klein's embeddings of $X(11)$ into $\mathbf{P}^4(\mathbf{C})$ as curves of degree 20 and 50. It is only a beginning, but I have high hopes for the method and have devoted considerable energy to developing some of the conceptual machinery and software tools required to make it viable.

Appendix IV is my unpublished article [A16]. It is devoted to the study of aspects of the geometry of the Hessian of a sufficiently generic cubic threefold Λ. One of these aspects is that the singular locus of the Hessian is a smooth curve of degree 20 and genus 26. In the special case of Klein's cubic threefold

$$(0.1) \qquad v^2w + w^2x + x^2y + y^2z + z^2v = 0,$$

the curve is isomorphic to the modular curve $X(11)$, as Klein himself discovered. Accordingly, we will call the singular locus of the Hessian of Λ the z-**curve of** Λ. This terminology is justified by the fact that it is likewise possible to generalize the A-curve to this context, as we will explain below. Most of the results of [A16] are taken from the unpublished article [A2], for example the explicit desingularization of the Hessian, the role played by Hecke operators in that desingularization and the theorem that the Jacobian variety of the z-curve of a sufficiently generic cubic threefold is the sum of an abelian variety $\mathcal{A}_{10}(\Lambda)$ of dimension 10 and one $\mathcal{A}_{16}(\Lambda)$ of dimension 16. In addition, there is an application of these results to the study of mirror symmetry and a discussion of the desingularization in connection with recent work on L^2 cohomology.

Appendix V, which is my unpublished article [A15], is devoted to the study of various new abelian varieties associated to a sufficiently generic cubic threefold Λ. Among these are the 10 and 16 dimensional factors of the jacobian of the z-curve of Λ. The first result of the paper is that the resulting abelian schemes over the space of all sufficiently generic cubic threefolds are simple. This result was sharpened by Torsten Ekedahl, who read the article as I was putting the finishing touches on the manuscript and who showed that in fact these abelian schemes are absolutely simple. His argument is sketched at the end of §46. In §47, I show that a sufficiently generic cubic is expressible in ∞^5 ways as the Pfaffian of a skew symmetric 6×6 matrix M whose entries are linear forms in 5 variables. For sufficiently generic M, the rank < 4 locus of M is empty in $\mathbf{P}^4$. That means that M gives rise to a rank 2 vector bundle $E(M)$ over Λ and that we obtain ∞^5 rank 2 vector bundles in this way. The projectivized vector bundle $\mathbf{P}(E(M))$ has a natural mapping into $\mathbf{P}^5$ and the image is a quartic hypersurface whose singular locus consists of a smooth irreducible curve. We call the curve **the A-curve of** M and **an A-curve of** Λ. This terminology is justified by the fact that in the case of Klein's cubic, one obtains Klein's A-curve of level 11 in this way, which incidently presents the equations of Klein's A-curve for the first time. We then show that the Jacobian variety of the A-curve of a sufficiently generic M decomposes into the sum of an abelian variety $\mathcal{A}_{21}$ of dimension 21 and one $\mathcal{A}_5$ of dimension 5. Furthermore, the 5 dimensional abelian variety is isogenous to the intermediate jacobian variety of the cubic threefold $\Lambda(M)$. I am indebted to Ofer Gabber, Yum-Tong Siu, Torsten Ekedahl and Gabor Megyesi for their help with the paper.

These two complementary appendices IV and V therefore serve to illustrate one of the advantages of a direct study of the equations of moduli spaces, namely that one can use them to discover new and unsuspected results about other varieties which at first glance seem to have nothing to do with the moduli spaces in question. The focus in these appendices has been on the modular curve $X(11)$ and its use in the study of cubic threefolds. But for other levels we can expect similar results and we expect the considerations of [A6] to play a role in obtaining them.

In a work that has taken twenty years from its origins in the paper [A2] in 1976-77 to the final form of the book in 1996, there are necessarily many people who deserve thanks. I am grateful to André Weil, who permitted me to visit the Institute for Advanced Study as his assistant during the years 1976-78 and who introduced me to the cubic threefold (0.1), the detailed study of which resulted in the seminal article [A2] which, in turn, stimulated this book and many of my other papers. I am grateful to W.L.Edge for his helpful correspondence over the last two decades and for his wonderful articles on classical geometry, including the article [E1] which so profoundly influenced my work. I am grateful to Mikhail Gromov for making it possible for me to visit   IHES in 1985, where the manuscript was first revised. I am happy to thank Roger Howe and Walter Feit for making it possible for me to use the facilities at Yale University during the academic year 1989-90, where I first typeset the article. I also thank Walter Strauss, Jeffrey Hoffstein of Brown University, John Montgomery of the University of Rhode Island, Professor Michel Broué of the Département de Mathématique et Informatique of the École Normale Supérieure in Paris, Constantine Tsinakis of Vanderbilt University and Jim Porter at Western Kentucky University for allowing me to use facilities at their institutions. I especially thank Mikhail Gromov and Marcel Berger for arranging for partial support from IHES from September 1993 to June 1994 and for officially giving me visitor status, with financial support, from July of 1994 to February of 1995. I also thank William Thurston and Lenore Blum and others for inviting me to MSRI in Berkeley for a portion of the academic year 1995-96 as a Visiting Professor. I am likewise indebted to Madhav Nori for inviting me to visit the University of Chicago in April, 1996.

I thank Linda Williams and Denys Duchier for many helpful electronic mail messages on how to improve my use of TEX. I also thank Anne-Marie Aubert for helpful correspondence regarding the Weil representation.

Finally, I would like to thank the people who provided me with food, shelter, loans and handouts during the period of my unemployment, which began in 1983 and from which I have yet to emerge.

Allan Adler
September, 1996

Additional comments and acknowledgments of the second author

I wish to thank the Institute for Advanced Studies for having me as a visitor during 1978-79 when substantial portions of the results were obtained.

I am immensely indebted to Allan for his efforts to see that this modest work got a physical existence at last. Indeed, tormented by the lop-sidedness of the relative effort put in, I offered Allan at one stage the option of publishing these results under his own name, but he handsomely refused. I thank him warmly for the many hours of intellectual stimulus he provided, not all of which was mathematical!

S.Ramanan
TIFR, Bombay, 1995

Chapter I

Standard Heisenberg Groups

§1   Heisenberg Groups

Let $\mathcal{S}$ denote the set of all finite sequences $(d_1, \ldots, d_g)$ of positive integers. If δ_1 and δ_2 are such sequences, denote by (δ_1, δ_2) the concatenation of δ_1 and δ_2. If n is a positive integer and δ is the sequence $(d_1, \ldots, d_g)$, write $n\delta$ to denote the sequence

$$n\delta = (nd_1, \ldots, nd_g).$$

For every positive integer d, write $K(d)$ denote the group

$$K(d) = \mathbf{Z}/d\mathbf{Z}$$

and $K(d)^*$ to denote the group

$$K(d)^* = Hom(K(d), k^\times)$$

of homomorphisms of $K(d)$ into $k^\times$. More generally, if $\delta = (\delta_1, \ldots, \delta_g)$ belongs to $\mathcal{S}$, write $K(\delta)$ to denote the group

$$K(\delta) = \overset{g}{\underset{i=1}{\oplus}} K(d_i)$$

and $K(\delta)^*$ to denote the group

$$K(\delta)^* = Hom(K(\delta), k^\times)$$

of all homomorphisms of $K(\delta)$ into $k^\times$; we can identify $K(\delta)^*$ with $\prod_{i=1}^{g} K(d_i)^*$. Write $H(\delta)$ to denote the group

$$H(\delta) = K(\delta) \times K(\delta)^*.$$

We define the group $G(\delta)$ to be the set

$$G(\delta) = k^\times \times H(\delta)$$

with the following multiplication law:

(1.1) $$(\alpha, x, l) \cdot (\alpha', x', l') = (\alpha\alpha' l'(x), \, x + x', \, ll')$$

The projection of $G(\delta)$ onto the factor $H(\delta)$ is a homomorphism which we will denote by ν. On the other hand, we will often regard $k^\times$ as a subgroup of $G(\delta)$ instead of introducing notation to denote the homomorphism $\alpha \mapsto (\alpha, 0, 1)$ of $k^\times$ into $G(\delta)$. In particular, we will freely multiply elements of $G(\delta)$ by elements of $k^\times$ without explicit comment.

§2 Representations of Heisenberg Groups

Denote by $V(\delta)$ the vector space

$$V(\delta) = Hom_{Ens}(K(\delta), k)$$

over k consisting of all functions f from $K(\delta)$ into k. For every $z = (\alpha, x, l)$ in $G(\delta)$, define the operator $\rho(z)$ on $V(\delta)$ by the rule

(2.1) $$[\,\rho(z)\,f\,](y) = \alpha \cdot l(y) \cdot f(x + y)$$

whenever f belongs to $V(\delta)$. When it is necessary to specify δ we will write $\rho_\delta(z)$ instead of $\rho(z)$. The mapping which associates to the element z of $G(\delta)$ the operator $\rho(z)$ is a representation of $G(\delta)$ on $V(\delta)$; we denote that representation by ρ_δ. It is, up to isomorphism, the only irreducible representation ρ of $G(\delta)$ on a vector space V of positive dimension over k such that for all α in $k^\times$ we have

$$\rho(\alpha, 0, 1) = \alpha \cdot 1_V.$$

The dimension of $V(\delta)$ will be denoted $|\delta|$ and is given by

$$|\delta| = \prod_{i=1}^{g} d_i.$$

If $\delta_1, \ldots, \delta_r$ are elements of $\mathcal{S}$ and if δ is the concatenation

$$\delta = (\delta_1, \ldots, \delta_r)$$

of $\delta_1, \ldots, \delta_r$, we will sometimes write

$$\begin{array}{ccc}
K(\delta_1, \ldots, \delta_r) & & K(\delta) \\
K(\delta_1, \ldots, \delta_r)^* & & K(\delta)^* \\
H(\delta_1, \ldots, \delta_r) & \text{respectively to denote} & H(\delta) \\
G(\delta_1, \ldots, \delta_r) & & G(\delta) \\
V(\delta_1, \ldots, \delta_r) & & V(\delta) \\
\rho_{(\delta_1, \ldots, \delta_r)} & & \rho_\delta
\end{array}$$

The same conventions will apply to other objects which we may associate to δ.

§3   Some Homomorphisms Between Abelian Groups

For every positive integer d, denote by

$$j_d : K(d) \to 2K(2d)$$

the unique isomorphism such that

$$j_d(1 + d\mathbf{Z}) = 2 + 2d\mathbf{Z}.$$

By means of the isomorphism j_d, we can identify the group $K(d)$ with the group $2K(2d)$. If $\delta = (d_1, \ldots, d_g)$ is an element of S, we have

$$K(\delta) = \overset{g}{\underset{i=1}{\oplus}} K(d_i)$$

and

$$K(2\delta)^* = \overset{g}{\underset{i=1}{\oplus}} K(2\,d_i)^*.$$

We can therefore form the sum

$$j_\delta = \oplus_{i=1}^g j_{d_i} : K(\delta) \to 2K(2\delta)$$

of the isomorphisms j_{d_i} for $i = 1, \ldots, g$ to obtain an isomorphism j_δ of $K(\delta)$ onto $2K(2\delta)$. We denote by

$$j_\delta^* : K(2\delta)^* \to K(\delta)^*$$

the homomorphism of $K(2\delta)^*$ onto $K(\delta)^*$, deduced by duality from the homomorphism of $K(\delta)$ into $K(2\delta)$ given by the composition

$$K(\delta) \xrightarrow{j_\delta} 2K(2\delta) \hookrightarrow K(2\delta).$$

The kernel of j_δ^* consists of all of the elements χ of $K(2\delta)^*$ such that $\chi^2 = 1$. It is the same as saying that the kernel of j_δ^* coincides with the kernel of the endomorphism of $K(2\delta)^*$ which associates to each element χ of $K(2\delta)^*$ the element χ^2. Therefore we can naturally identify $K(\delta)^*$ with the group of all characters of $K(2\delta)$ which are of the form χ^2 for some character χ of $K(2\delta)$. If ψ is an element of $K(\delta)^*$, the corresponding element χ^2 of $K(2\delta)^*$ will be denoted by $_*j_\delta(\psi)$. If ψ is such and $\chi^2 = {_*j_\delta(\psi)}$, then for all x in $K(2\delta)$ we have

$$(3.1) \qquad {_*j_\delta(\psi)(x)} = \chi^2(x) = \chi(2x) = (j_\delta^*\chi)(j_\delta^{-1}(2x)) = \psi \circ j_\delta^{-1}(2x)$$

For future reference, we note that if $\lambda = {_*j_\delta(\psi)}$ and $x \in K(2\delta)$, we have

$$(3.2) \qquad \lambda(x) = ({_*j_\delta^{-1}(\lambda)})(j_\delta^{-1}(2x)).$$

§4   Some Homomorphisms Between Heisenberg Groups

Let δ be an element of S and let (x, l) be an element of $H(2\delta)$ such that

$$2x = 0$$

and

$$l^2 = 1$$

Denote by

$$\eta_{(x,l)} : G(2\delta) \to G(\delta)$$

the function from $G(2\delta)$ to $G(\delta)$ given by

(4.1) $$\eta_{(x,l)}(\alpha, y, m) = \left(\alpha^2 \cdot l(y) \cdot m(x),\, j_\delta^{-1}(2y),\, _*j_\delta^{-1}(m^2)\right)$$

It is straightforward to show that $\eta_{(x,l)}$ is a homomorphism of $G(2\delta)$ onto $G(\delta)$. Furthermore, one can show that the kernel of $\eta_{(x,l)}$ is the set

$$ker\ \eta_{(x,l)} = \left\{z = (\alpha, y, m) \in G(2\delta) \mid z^2 = (m(y)\ell(y)m(x), 0, 1)\right\}$$

of all $z = (\alpha, y, m)$ in $G(2\delta)$ such that $z^2 = (m(y)\ell(y)m(x), 0, 1)$. The following lemma is then immediate.

Lemma (4.2):   *Denote by $Q_{(x,l)}$ the mapping*

$$Q_{(x,l)} : H(2\delta) \to k^\times$$

given by

$$Q_{(x,l)}(y, m) = m(y) \cdot l(y) \cdot m(x).$$

Then the kernel of $\eta_{(x,l)}$ consists of all z in $G(2\delta)$ such that

$$z^2 = Q_{(x,l)}(\nu(z)) \in k^\times \subset G(\delta).$$

Denote by

$$D_\delta : G(\delta) \to G(\delta)$$

the automorphism of $G(\delta)$ given by

(4.3) $$D_\delta(\alpha, x, l) = (\alpha, -x, l^{-1}).$$

We will denote by $Aut(\delta)$ the group of all automorphisms of $G(\delta)$ and by $Aut_0(\delta)$ the subgroup of $Aut(\delta)$ consisting of all automorphisms which induce the identity mapping on the center $k^\times$ of $G(\delta)$. The subgroups of $Aut(\delta)$ and $Aut_0(\delta)$ consisting of automorphisms which commute with the automorphism D_δ will be denoted $Aut^D(\delta)$ and $Aut_0^D(\delta)$ respectively. We will call the elements of $Aut_0^D(\delta)$ **symmetric automorphisms** of $G(\delta)$.

§5   The Homomorphism $T_{(x,l)}$

The δ be an element of $\mathcal{S}$ and let (x, l) be an element of $H(2\delta)$ such that $2x = 0$ and $l^2 = 1$. Denote by $G(2\delta, 2\delta)'$ the subgroup

$$G(2\delta, 2\delta)' = \{z = (\alpha, x_1, x_2, l_1, l_2) \in G(2\delta, 2\delta) \mid x_1 + x_2 \in im(j_\delta),\ l_1 l_2 \in im(_* j_\delta)\}$$

of $G(2\delta, 2\delta)$ consisting of all $(\alpha, x_1, x_2, l_1, l_2)$ such that $x_1 + x_2$ lies in the image of j_δ and $l_1 l_2$ lies in the image of $_* j_\delta$. Denote by

$$T_{(x,l)} : G(2\delta, 2\delta)' \to G(\delta, \delta)$$

the function from $G(2\delta, 2\delta)'$ to $G(\delta, \delta)$ given by

$$(5.1)\qquad T_{(x,l)}(\alpha, x_1, x_2, l_1, l_2) = (\alpha \cdot l(x_1) \cdot l_1(x),\ y_1,\ y_2,\ m_1,\ m_2)$$

where the elements y_1, y_2 of $K(\delta)$ and the elements m_1, m_2 of $K(\delta)^*$ are determined by the equations

$$j_\delta(y_1) = x_1 + x_2 \quad j_\delta(y_2) = x_1 - x_2$$

(5.2)

$$_* j_\delta(m_1) = l_1 l_2 \quad _* j_\delta(m_2) = l_1 l_2^{-1}$$

Proposition (5.2):   *The mapping $T_{(x,l)} : G(2\delta, 2\delta)' \to G(\delta, \delta)$ given by (5.1) is a surjective homomorphism whose kernel consists of all elements of $G(2\delta, 2\delta)'$ which can be written in the form*

$$(l(u) \cdot \chi(x),\, u, u, \chi, \chi)$$

where (u, χ) is an element of $H(\delta)$ such that $2u = 0$ and $\chi^2 = 1$.

Proof: It is clear that:

(1) the restriction of $T_{(x,l)}$ to $k^\times$ is a homomorphism;
(2) $T_{(x,l)}$ preserves cosets of $k^\times$;
(3) $T_{(x,l)}$ induces a homomorphism from $G(2\delta, 2\delta)'/k^\times$ to $G(\delta, \delta)/k^\times$.

Therefore, we will be done if we can show that

$$T_{(x,l)}(1, x_1, x_2, 1, 1) \cdot T_{(x,l)}(1, 0, 0, l_1, l_2) = T_{(x,l)}(l_1(x_1)l_2(x_2), x_1, x_2, l_1, l_2).$$

Expanding the two sides, this amounts to showing that

$$m_1(y_1)m_2(y_2) = l_1(x_1)l_2(x_2).$$

However

$$m_1(y_1)m_2(y_2) = (_* j_\delta^{-1}(l_1 l_2))(j_\delta^{-1}(x_1 + x_2))(_* j_\delta^{-1}(l_1 l_2^{-1}))(j_\delta^{-1}(x_1 - x_2))$$
$$= (_* j_\delta^{-1}(l_1 l_2))(j_\delta^{-1}(2z_1))(_* j_\delta^{-1}(l_1 l_2^{-1}))(j_\delta^{-1}(2z_2)),$$

where z_1 is an element of $K(2\delta)$ such that $2z_1 = x_1 + x_2$ and where $z_2 = x_1 - z_1$. Applying (3.2) this becomes

$$(l_1 l_2)(z_1)(l_1 l_2^{-1})(z_2) = l_1(z_1 + z_2)l_2(z_1 - z_2) = l_1(x_1)l_2(x_2)$$

as required. The determination of the kernel of $T_{(x,1)}$ is straightforward and is left to the reader.

§6   The Linear Operator $\Omega_{(x,1)}$

We will retain the notation of the preceding paragraph. Denote by

$$\Omega_{(x,1)} : V(\delta, \delta) \to V(2\delta, 2\delta)$$

the linear mapping of $V(\delta, \delta)$ into $V(2\delta, 2\delta)$ which associates to a function Φ on $K(\delta, \delta)$ the function Ψ on $K(2\delta, 2\delta)$ defined by
(6.1)

$$\Psi(v, w) = \begin{cases} l(v) \cdot \Phi\left(j_\delta^{-1}(v + w + x), \, j_\delta^{-1}(v - w + x)\right) & \text{if } v + w + x \in im(j_\delta) \\ 0 & \text{if not} \end{cases}$$

It is quite easy to show that any function Ψ on $K(\delta, \delta)$ obtained in this manner is fixed by the operator $\rho_{(2\delta, 2\delta)}(z)$ whenever z lies in the kernel of $T_{(x,1)}$. In fact, let us write z in the form

$$z = (l(u)\chi(x), u, u, \chi, \chi)$$

where (u, χ) is an element of $H(2\delta)$ such that $2u = 0$ and $\chi^2 = 1$, and write Ψ' to denote the image of Ψ under $\rho_{(2\delta, 2\delta)}(z)$. If (v, w) is an element of $K(2\delta, 2\delta)$ such that $v + w + x$ lies in the image of j_δ, say $v + w + x = j_\delta(a)$, then we have

(6.2) $$\chi(v)\chi(w)\chi(x) = \chi(v + w + x) = 1,$$

since $\chi^2 = 1$ and since $2K(2\delta)$ is the image of j_δ. Therefore, since Ψ is supported on the elements (v, w) of $K(2\delta, 2\delta)$ such that $v + w + x$ lies in the image of j_δ, we have

$$(\rho_{(2\delta, 2\delta)}(z)\Psi)(v, w) = l(u)\Psi(u + v, u + w)$$

(6.3)
$$= \begin{cases} l(v)\Phi(j_\delta^{-1}(V), j_\delta^{-1}(W)) & \text{if } V \in im(j_\delta) \\ 0 & \text{if not.} \end{cases}$$

$$= \Psi(v, w)$$

where $V = v + w + x$ and $W = v - w + x$ and $im(j_\delta)$ denotes the image of j_δ. This proves that $\Omega_{(x,1)}\Phi$ is fixed by $\rho_{(2\delta, 2\delta)}(z)$ for all z in the kernel of $T_{(x,1)}$.

Proposition (6.4): *The image of the linear mapping*

$$\Omega_{(x,1)} : V(\delta, \delta) \to V(2\delta, 2\delta)$$

consists of all functions Ψ *in* $K(2\delta, 2\delta)$ *which are fixed by every operator of the form* $\rho_{(2\delta,2\delta)}(z)$ *with* z *in the kernel of* $T_{(x,l)}$.

Proof: Let X denote the image of $\Omega_{(x,l)}$ and let Y denote the space of all functions Ψ on $K(2\delta, 2\delta)$ which are fixed by every operator of the form $\rho_{(2\delta,2\delta)}(z)$ with z in the kernel of $T_{(x,l)}$. To simplify notation, if

$$z = (\, l(u) \cdot \chi(x)\,, u, u, \chi, \chi)$$

is any element of the kernel of $T_{(x,l)}$, write $A(u, \chi)$ to denote the operator $\rho_{(2\delta,2\delta)}(z)$. A function Ψ on $K(2\delta, 2\delta)$ is fixed by $A(u, \chi)$ if and only if for all (v, w) in $K(2\delta, 2\delta)$ we have

(6.5) $\Psi(v, w) = (\, A(u, \chi)\, \Psi\,)(v, w) = l(u) \cdot \chi(x + v + w) \cdot \Psi(u + v, u + w)$

Taking $u = 0$ in (6.5) and letting χ run over all characters of $K(2\delta)$ such that $\chi^2 = 1$, we have that Ψ must be supported on the set of all (v, w) such that $v + w + x$ lies in $2K(2\delta)$ or, what is the same, in the image of j_δ. On the other hand, if we let $\chi = 1$ in (6.5), we see that the value of $\Psi(u + v, u + w)$ is deducible from the value of $\Psi(v, w)$ whenever u is an element of $K(2\delta)$ such that $2u = 0$. Consequently the dimension of the space Y is not greater than the number of cosets of the form $(v, w) + \Delta$ such that $v + w + x$ is in the image of j_δ, where we have written Δ to denote the group of all elements of $K(2\delta, 2\delta)$ of the form (u, u) with $2u = 0$. The dimension of Y is therefore not greater than $|\delta|^2 = |(\delta, \delta)|$, which is also the dimension of $V(\delta, \delta)$. Since $\Omega_{(x,l)}$ is clearly injective and since we have already shown that X is contained in Y, it follows that X equals Y.

§7 $\Omega_{(x,l)}$ as an Intertwining Operator

In this section we will prove that linear mapping $\Omega_{(x,l)}$ defined in the preceding section has certain invariance properties which will be essential to us in proving the addition formula in §17.

Proposition (7.1):   *Let* z *be an element of* $G(2\delta, 2\delta)'$ *and write* A *to denote the operator* $\rho_{(2\delta,2\delta)}(z)$ *on* $V(2\delta, 2\delta)$. *Let* B *denote the operator* $\rho_{(\delta,\delta)}\big(T_{(x,l)}(z)\big)$ *on* $V(\delta, \delta)$. *Then we have*

(7.2) $\Omega_{(x,l)} \circ B = A \circ \Omega_{(x,l)}.$

Proof: We can write z in the form $(\alpha, x_1, x_2, l_1, l_2)$. Then we can find (y_1, y_2, m_1, m_2) in $H(\delta, \delta)$ such that

$$j_\delta(y_1) = x_1 + x_2, \quad j_\delta(y_2) = x_1 - x_2$$

(7.3)

$$_*j_\delta(m_1) = l_1 l_2, \quad _*j_\delta(m_2) = l_1 l_2^{-1}.$$

Let Φ be an element of $V(\delta, \delta)$. Let

$$\Psi = \Omega_{(x,l)}(\Phi)$$

and

$$\Psi' = \Omega_{(x,l)} \circ B(\Phi).$$

Then we have for all u, v in $K(2\delta)$ that

$$(A\Psi)(u, v) = \alpha l_1(u) l_2(v) \Psi(u + x_1, v + x_2)$$

(7.4)
$$= \begin{cases} \alpha l_1(u) l_2(v) l(u + x_1) \Phi(j_\delta^{-1}(U) + y_1, j_\delta^{-1}(V) + y_2) & \text{if } U \in im(j_\delta) \\ 0 & \text{if not} \end{cases}$$

where

$$U = u + v + x$$

and

$$V = u - v + x$$

and where $im(j_\delta)$ denotes the image of j_δ. Since

$$T_{(x,l)}(z) = (\alpha l(x_1) \cdot l_1(x), y_1, y_2, m_1, m_2)$$

we have for all a, b in $K(\delta)$ that

(7.5) $$(B\Phi)(a, b) = \alpha l(x_1) \cdot l_1(x) \cdot m_1(a) \cdot m_2(b) \cdot \Phi(a + y_1, b + y_2).$$

Since $\Psi' = \Omega_{(x,l)} \circ B(\Phi)$, we therefore have for all u, v in $K(\delta)$ that $\Psi'(u, v) = 0$ if U does not lie in the image of j_δ and that
(7.6)
$$\Psi'(u, v) = \alpha l(u) \cdot B\Phi\left(j_\delta^{-1}(U), j_\delta^{-1}(V)\right)$$

$$= \alpha l(u + x_1) \cdot l_1(x) \cdot m_1(j_\delta^{-1}(U)) \cdot m_2(j_\delta^{-1}(V)) \cdot \Phi\left(j_\delta^{-1}(U) + y_1, j_\delta^{-1}(V) + y_2\right)$$

if U does lie in the image of j_δ. The proposition will be proved if we show that

$$A\Psi = \Psi'.$$

This in turn will follow if we can show that

(7.7) $$l_1(u) \cdot l_2(v) \cdot l(u + x_1) = l(u + x_1) \cdot l_1(x) \cdot m_1(j_\delta^{-1}(U)) \cdot m_2(j_\delta^{-1}(V))$$

whenever u, v are elements of $K(2\delta)$ such that $u + v + x$ lies in the image of j_δ. We can cancel $l(u + x_1)$ from both sides, so it is enough to prove

(7.8) $$l_1(u) \cdot l_2(v) = l_1(x) \cdot m_1\left(j_\delta^{-1}(U)\right) \cdot m_2\left(j_\delta^{-1}(V)\right).$$

Since $2K(2\delta)$ is the image of j_δ, we can write $U = 2u'$ with $u' \in K(\delta)$. If we then let $v' = u - u' + x$, we have

$$u = u' + v' + x$$

and

$$v = u' = v'$$

with u' and v' in $K(2\delta)$. Then we have

$$(7.9) \quad l_1(x) \cdot m_1\left(j_\delta^{-1}(U)\right) \cdot m_2\left(j_\delta^{-1}(V)\right) = l_1(x) \cdot m_1\left(j_\delta^{-1}(2u')\right) \cdot m_2\left(j_\delta^{-1}(2v')\right).$$

If we now apply equations (7.3) and (3.2), the right hand side becomes

$$
\begin{aligned}
l_1(x) \cdot [(l_1 l_2)(u')][l_1 l_2^{-1}(v')] \quad &= \quad l_1(x) \cdot l_1(u' + v') \cdot l_2(u' - v') \\
(7.10) \qquad\qquad &= \quad l_1(x) \cdot l_1(u + x) \cdot l_2(v) \\
&= \quad l_1(u) \cdot l_2(v).
\end{aligned}
$$

Let V denote the subspace of $V(2\delta, 2\delta)$ consisting of all functions Ψ on $K(2\delta, 2\delta)$ which are fixed by every operator of the form $\rho_{(2\delta,2\delta)}(z)$ with z in the kernel of $T_{(x,l)}$. Then V is invariant under all of the operators of the form $\rho_{(2\delta,2\delta)}(w)$ with w in $G(2\delta, 2\delta)'$. We can therefore regard V as a module for the factor group $G(2\delta, 2\delta)'/K$ where K is the kernel of $T_{(x,l)}$. Since $T_{(x,l)}$ maps $G(2\delta, 2\delta)'$ onto $G(\delta,\delta)$, we can use the natural isomorphism of $G(2\delta, 2\delta)'/K$ onto $G(\delta,\delta)$ induced by $T_{(x,l)}$ to regard V as a module for $G(\delta,\delta)$.

Corollary (7.11):   The linear mapping $\Omega_{(x,l)}$ is an isomorphism of the $G(\delta,\delta)$-module $V(\delta,\delta)$ onto the $G(\delta,\delta)$-module V.

Proof: We know from Proposition (6.4) that $\Omega_{(x,l)}$ maps the vector space $V(\delta,\delta)$ isomorphically onto the vector space V. By Proposition (7.1), $\Omega_{(x,l)}$ is a morphism of $G(\delta,\delta)$-modules.

§8   Characterization of the Operator $T_{(x,l)}$

In this section we will give a useful characterization of the homomorphism $T_{(x,l)}$ which was defined in §5. Our characterization will be phrased in terms of certain commutative diagrams. So we begin by defining some of the morphisms which appear in these diagrams. As in §5, (x, l) will denote an element of $H(2\delta)$ such that $2x = 0$ and $l^2 = 1$.

Define the homomorphisms S_1 and S_2 from $G(\delta)$ to $G(\delta, \delta)$ by the rules

$$S_1(\alpha, y, m) = (\alpha, y, 0, m, 1)$$

(8.1)

$$S_2(\alpha, y, m) = (\alpha, 0, y, 1, m)$$

We also define homomorphisms A and A' from $G(2\delta)$ to $G(2\delta, 2\delta)'$ by the rules

(8.2)
$$A(\alpha, y, m) = (\alpha^2, y, y, m, m)$$

$$A'(\alpha, y, m) = (\alpha^2, y, -y, m, m^{-1})$$

It is then straightforward to verify that the diagrams (8.3) and (8.4) given below are commutative.

(8.3)
$$
\begin{array}{ccc}
G(2\delta) & \overset{\eta_{(x,l)}}{\longrightarrow} & G(\delta) \\
\Big\downarrow A & & \Big\downarrow S_1 \\
G(2\delta, 2\delta)' & \overset{T_{(x,l)}}{\longrightarrow} & G(\delta, \delta)
\end{array}
$$

(8.4)
$$
\begin{array}{ccc}
G(2\delta) & \overset{\eta_{(x,l)}}{\longrightarrow} & G(\delta) \\
\Big\downarrow A' & & \Big\downarrow S_2 \\
G(2\delta, 2\delta)' & \overset{T_{(x,l)}}{\longrightarrow} & G(\delta, \delta)
\end{array}
$$

Here $\eta_{(x,l)}$ denotes the homomorphism of $G(2\delta)$ onto $G(\delta)$ defined by equation (4.1). It can be easily verified that the group $G(2\delta, 2\delta)'$ is generated by the image of A and the image of A'. It follows that the commutative diagrams (8.3) and (8.4) determine the homomorphism $T_{(x,l)}$ uniquely.

Chapter II

Heisenberg Groups of Line Bundles on Abelian Varieties

§9   Heisenberg Groups of Sheaves on Abelian Varieties

If X is an abelian variety defined over k and x is an element of X, denote by T_x the morphism

$$T_x : X \to X$$

from X to itself given by

$$T_x(y) = x + y.$$

If L is any sheaf on X, denote by $\Gamma(L)$ the group of sections of L and by $H(L)$ the subgroup of X consisting of all x in X such that $T_x^* L$ is isomorphic to L. Denote by $G(L)$ the set

$$G(L) = \left\{ (x, \varphi) \mid x \in H(L), \ \phi : L \xrightarrow{\sim} T_x^* L \right\}$$

of all pairs (x, φ) in which x is an element of $H(L)$ and φ is an isomorphism of L onto $T_x^*(L)$. We can make $G(L)$ into a group whose operation is given by

$$(9.1) \qquad\qquad (x, \phi) \cdot (y, \psi) = \left(x + y, \, (T_x^* \psi) \circ \phi \right).$$

We then have an exact sequence

$$(9.2) \qquad\qquad 1 \to Aut(L) \to G(L) \to H(L) \to 0$$

where $Aut(L)$ denotes the automorphism group of the sheaf L. We will call $G(L)$ the **Heisenberg group** of L and we will denote by ν_L or ν the homomorphism from $G(L)$ to $H(L)$ given by $(x, \varphi) \mapsto x$. In case L is an invertible sheaf, we can naturally identify $Aut(L)$ with $k^\times$. Furthermore, if L is ample then $H(L)$ will be a finite group. **We will assume in the rest of this section that L is an ample invertible sheaf which we will assume to be of separable type.** The latter assumption means that the natural morphism associated to L of X into the Picard variety of X is separable. One can then show (cf. Corollary of Theorem 1 of [Mul], p.294) that there is an element $\delta = (d_1, \ldots, d_g)$ of S and an isomorphism θ of $G(L)$ onto $G(\delta)$ which is the identity on $k^\times$ and such that $g = dim\ X$.

The group $G(L)$ acts in a natural way on the space $\Gamma(L)$ of all sections of L. Namely, if $z = (x, \varphi)$ belongs to $G(L)$ and if σ is a section of L, then let $\rho_L(z)(\sigma)$ be the section

$$\rho_L(z)(\sigma) = T_{-x}^*(\varphi \circ \sigma)$$

of L. The mapping ρ_L which associates to each z in $G(L)$ the operator $\rho_L(z)$ is a representation of $G(L)$ on $\Gamma(L)$. If we choose an isomorphism θ of $G(L)$ onto $G(\delta)$ as above, then we can view $\Gamma(L)$ as a $G(\delta)$-module. It is easy to see that for t in $k^\times$, we have

$$\rho_L(t) = t \circ 1_{\Gamma(L)}.$$

Furthermore. one can show that the dimension of $\Gamma(L)$ is equal to $|\delta|$. It follows then from the characterization in §1 of the irreducible representation ρ_δ that $\Gamma(L)$ is isomorphic as a $G(\delta)$-module to $V(\delta)$. In particular, the representation ρ_L is irreducible.

Suppose Y is another abelian variety, M is an ample invertible sheaf of separable type on Y and $f : X \to Y$ is a separable morphism with finite kernel such that f^*M is isomorphic to L. Let

$$\alpha : L \to f^*M$$

be an isomorphism. Let $G(L)'$ denote the subgroup

$$G(L)' = \{(x, \varphi) \in G(L) \mid f(x) \in H(M)\}$$

of $G(L)$ consisting of all pairs (x, φ) such that $f(x)$ lies in $H(M)$. If (x, φ) belongs to $G(L)'$, there is one and only one isomorphism

$$\psi : M \to T_y^*(M),$$

where $y = f(x)$, such that

$$(f^*\psi) \circ \alpha = (T^*\alpha) \circ \varphi.$$

The mapping

$$h_{(f,\alpha)} : G(L)' \to G(M)$$

of $G(L)'$ into $G(M)$ which associates to (x, φ) the pair (y, ψ) is a homomorphism. If it is necessary to specify the morphism f when speaking of $G(L)'$, we will write $G(L)'_{(f,\alpha)}$ instead of $G(L)'$.

§10   The Mumford Functor

The systems of groups, representations and morphisms which we have associated in the preceding section to ample invertible sheaves of separable type have certain functoriality properties which we will formulate in this section. We denote by $\mathcal{A}$ the category whose objects are all pairs (X, L) consisting of an abelian variety X and an ample invertible sheaf L of separable type on X. If (X, L) and (Y, M) are objects of $\mathcal{A}$, a morphism from (X, L) to (Y, M) will be a pair (f, α) consisting of a morphism $f : X \to Y$ of varieties and an isomorphism α of L onto f^*M. We will refer to $\mathcal{A}$ as the **category of separably polarized abelian varieties**. .

We will denote by $Heis$ the category whose objects are pairs (G, ρ) where G is a group and ρ is a finite dimensional representation of G on a vector space V_ρ. If (G, ρ) and (Γ, R) are objects of the category $Heis$, then a morphism from (G, ρ) to

(Γ, R) will be a triple (G', h, Ω) in which

(i) G' is a subgroup of finite index in G;

(ii) h is a homomorphism from G' into Γ whose image is a subgroup of finite index in Γ;

(iii) Ω is a linear mapping of V_R into V_ρ such that for all g' in G' we have

(10.1) $$\rho(g') \circ \Omega = \Omega \circ R(g)$$

where $g = h(g')$.

Suppose (G', h, Ω) is a morphism from (G, ρ) to (Γ, R) and (Γ', h', Ω') is a morphism from (Γ, R) to $(\mathcal{G}, \mathcal{R})$. Then we define the **composition** of (G', h, Ω) and (Γ', h', Ω') to be the morphism (G'', h'', Ω'') from (G, ρ) to $(\mathcal{G}, \mathcal{R})$ satisfying (a), (b) and (c) below:

(a) G'' is the preimage of Γ' under h.

(b) h'' is the composition of h' with the restriction of h to G'', i.e.
$h'' = h' \circ (h|G'')$.

(c) $\Omega'' = \Omega \circ \Omega'$.

With the objects, morphisms and composition of morphisms so defined, *Heis* now becomes a category which we call the **Heisenberg category**.

We now define a functor, which we call the **Mumford functor** and denote by *Mum*, from the category $\mathcal{A}$ to the category *Heis*. If (X, L) is an object of $\mathcal{A}$ we define $Mum((X, L))$ to be the object $(G(L), \rho_L)$ of *Heis*. If $(f, \alpha) : (X, L) \to (Y, M)$ is a morphism of $\mathcal{A}$, we define $Mum((f, \alpha))$ to be the morphism from $(G(L), \rho_L)$ to $(G(M), \rho_M)$ given by

(10.2) $$Mum(f, \alpha) = (G(L)'_{(f,\alpha)}, h_{(f,\alpha)}, \Omega_{(f,\alpha)})$$

where $G(L)'_{(f,\alpha)}$ and $h_{(f,\alpha)}$ are as defined in the preceding section and where $\Omega_{(f,\alpha)}$ is the linear mapping of $\Gamma(M)$ to $\Gamma(L)$ given by

(10.3) $$\Omega_{(f,\alpha)}(\sigma) = \alpha^{-1} \circ f^*(\sigma).$$

The verification that *Mum* is a functor is entirely straightforward and is left to the reader.

§11   Strongly Symmetric Line Bundles

We will use the Mumford functor in §13 to show that certain diagrams in the Heisenberg category are commutative. In order to do that, we first describe some morphisms and commutative diagrams in the category $\mathcal{A}$.

For every abelian variety X and every integer n, denote by

$$n_X : X \to X$$

the endomorphism of X given by

$$n_X(x) = n \cdot x$$

for all elements of X. For $n = 1$, this coincides with the usual notation 1_X for the identity morphism of X. We will also write ι to denote -1_X. The kernel of n_X is denoted X_n. The diagonal morphism of X into $X \times X$ is denoted Δ and is defined by

$$\Delta(x) = (x, x).$$

On the other hand, we will write Δ' to denote the "antidiagonal" morphism

$$\Delta' = (1_X, \iota) \circ \Delta : X \to X \times X$$

of X into $X \times X$. It is defined by

$$\Delta'(x) = (x, -x).$$

Denote by s_1 and s_2 the morphisms of factor inclusion

$$s_1, s_2 : X \to X \times X$$

of X into $X \times X$ given by

$$s_1(x) = (x, 0)$$

and

$$s_2(x) = (0, x)$$

respectively. We will also need the endomorphism $\xi = \xi_X$ of $X \times X$ given by

$$\xi(x, y) = (x + y, x - y).$$

For typographical reasons we will usually omit the subscript X except in cases where it is important to specify the abelian variety. The projections of $X \times X$ onto its first and second factors will be denoted p_1 and p_2 respectively. We summarize below the maps we have just defined.

(11.1)
$$
\begin{aligned}
\Delta(x) &= (x, x) \\
\Delta'(x) &= (x, -x) \\
s_1(x) &= (x, 0) \\
s_2(x) &= (0, x) \\
\xi(x, y) &= (x + y, x - y) \\
p_1(x, y) &= x \\
p_2(x, y) &= y
\end{aligned}
$$

If L_1 and L_2 are sheaves of $\mathcal{O}_X$-modules on X, we denote by $L_1 \boxtimes L_2$ the exterior tensor product sheaf of $\mathcal{O}_{X \times X}$-modules on $X \times X$ defined by

(11.2) $$L_1 \boxtimes L_2 = p_1^*(L_1) \otimes p_2^*(L_2).$$

Definition (11.3):   *Let X be an abelian variety and let L be a sheaf on X. We say that L is **symmetric** if ι^*L is isomorphic to L. If (X, L) is an object of the category $\mathcal{A}$ and if L is symmetric, we will say that (X, L) is symmetric.*

Let (X, L) be a symmetric object of the category $\mathcal{A}$. There is one and only one isomorphism ψ of L onto ι^*L such that ψ induces the identity morphism in the stalk of L at 0. It follows easily that $\psi \circ \iota^*\psi = 1_L$. We will call ψ the **normalized isomorphism** of L onto ι^*L. For every x in X, denote by L_x the stalk of L at x. Then for every x in X_2, ψ induces an endomorphism ψ_x of L_x. The endomorphism ψ_x must have the form $\epsilon(x) \cdot 1_M$, where $M = L_x$ and where $\epsilon(x) = \pm 1$ for all x in X_2. One can show that the mapping

$$\epsilon : X_2 \to k^\times$$

of X_2 into $k^\times$ which associates to each x in X_2 the value $\epsilon(x)$ is a **quadratic character**. It is the same to say that for all x and y in X_2. the pairing

$$\beta : X_2 \times X_2 \to k^\times$$

of $X_2 \times X_2$ into $k^\times$ given by

$$(11.4) \qquad\qquad \beta(x, y) = \epsilon(x + y) \cdot \epsilon(x)^{-1} \cdot \epsilon(y)^{-1}$$

is a homomorphism in each variable separately; that is to say, β is a **bicharacter** of $X_2 \times X_2$. When it is important to specify L, we will write ϵ_L instead of ϵ.

Definition (11.5):   Let (X, L) be a symmetric object of $\mathcal{A}$. We will say that (X, L) is **strongly symmetric** and that L is **strongly symmetric** in case $\epsilon(x) = 1$ for every x which lies in the intersection of X_2 and $H(L)$.

Remark (11.6):   If $H(L)$ is of odd order, then there is no difference between symmetry and strong symmetry. On the other hand, if L is the square of a line bundle (which is the case if and only if $X_2 \subset H(L)$), then strong symmetry is equivalent to total symmetry as defined in [Mu1, I].

If L is any invertible sheaf on X and n is an integer, then n^*L is isomorphic to $L^a \otimes \iota^*L^b$ where

$$a = \frac{1}{2}n(n + 1)$$

and

$$b = \frac{1}{2}n(n - 1).$$

In case L is symmetric, n^*L is isomorphic to L^c, where $c = a + b = n^2$, although in general there is no canonical way to choose such an isomorphism. Let M denote the sheaf $L \boxtimes L$ on $X \times X$. Then

$$s_1^*M = s_2^*M = L.$$

One of the basic facts of the theory is that if we suppose that L is symmetric, then

$\xi^* M$ is isomorphic to M^2. Finally, we have that $\Delta^* M^2$ is isomorphic to L^4.

§12   Some Commutative Diagrams in the Category $\mathcal{A}$

We will retain the notation and assumptions of the preceding section. We begin with the two commutative diagrams given below.

(12.1)
$$
\begin{array}{ccc}
X & \xrightarrow{\;2\;} & X \\
\downarrow{\scriptstyle \Delta} & & \downarrow{\scriptstyle s_1} \\
X \times X & \xrightarrow{\;\xi\;} & X \times X
\end{array}
$$

(12.2)
$$
\begin{array}{ccc}
X & \xrightarrow{\;2\;} & X \\
\downarrow{\scriptstyle \Delta'} & & \downarrow{\scriptstyle s_2} \\
X \times X & \xrightarrow{\;\xi\;} & X \times X
\end{array}
$$

Then if we choose isomorphisms

(12.3)
$$
\begin{aligned}
\alpha : L^4 &\rightarrow 2^* L \\
\beta : M^2 &\rightarrow \xi^* M \\
\gamma : L^4 &\rightarrow \Delta'^* M^2
\end{aligned}
$$

we can form the following two diagrams in the category $\mathcal{A}$, but we do not claim the diagrams are commutative.

(12.4)
$$
\begin{array}{ccc}
(X, L^4) & \xrightarrow{(2,\alpha)} & (X, L) \\
\downarrow{\scriptstyle (\Delta, id)} & & \downarrow{\scriptstyle (s_1, id)} \\
(X \times X, M^2) & \xrightarrow{(\xi, \beta)} & (X \times X, M)
\end{array}
$$

(12.5)
$$
\begin{array}{ccc}
(X, L^4) & \xrightarrow{(2,\alpha)} & (X, L) \\
\downarrow{\scriptstyle (\Delta', \gamma)} & & \downarrow{\scriptstyle (s_2, id)} \\
(X \times X, M^2) & \xrightarrow{(\xi, \beta)} & (X \times X, M)
\end{array}
$$

In fact, there is no reason to suppose that these diagrams commute. However, in diagram (12.4), the two morphisms $(X, L^4) \rightarrow (X \times X, M)$ obtained by composing along the high road and the low road respectively will be of the form $(s_1 \circ 2, \alpha')$ and $(s_1 \circ 2, \beta')$, where α' and β' are isomorphisms of L^4 onto $(s_1 \circ 2)^* M$. Therefore α' and β' differ only by a nonzero scalar factor c. If we replace α by $c^{-1}\alpha$ then the diagram (12.4) will commute. Once we have fixed α, we can then modify γ by a suitable scalar factor so that the diagram (12.5) is also commutative. We record these observations in the following lemma.

Lemma (12.6):   Let X be an abelian variety. Let L be an ample symmetric invertible sheaf of separable type on X and let M denote the sheaf $L \boxtimes L$ on $X \times X$. Let $2 = 2_X$, $\xi = \xi_X$, $\Delta = \Delta_X$, $\Delta' = \Delta'_X$, s_1 and s_2 be as in §11. Then the diagrams (12.1) and (12.2) are commutative. Furthermore, there are isomorphisms α, β and

γ as in (12.3) such that the diagrams (12.4) and (12.5) are commutative.

§13   Application of the Mumford functor to the Preceding Diagrams

We will retain the notation of the preceding section. In particular, X is an abelian variety and L is an ample symmetric line bundle of separable type on X. If (x, ϕ) is an element of $G(L^2)$, then the isomorphism ϕ of L^2 onto $T_x^* L^2$ induces an isomorphism of L^4 onto $T_x^* L^4$; that isomorphism will be denoted $\phi^{\otimes 2}$. The pair $(x, \phi^{\otimes 2})$ is therefore an element of $G(L^4)$. The mapping

$$E : G(L^2) \to G(L^4)$$

of $G(L^2)$ into $G(L^4)$ given by

$$E(x, \phi) = (x, \phi^{\otimes 2})$$

is a homomorphism.

  Let α be the isomorphism of L^4 onto $2_X^* L$ given in Lemma (12.6). Applying the Mumford functor to the morphism $(2_X, \alpha)$ of (X, L^4) into (X, L) we obtain, among other things, a homomorphism $h_{(2,\alpha)}$ of $G(L^4)'_{(2,\alpha)}$ into $G(L)$. By definition (cf.§9), $G(L^4)'_{(2,\alpha)}$ consists of all (x, ϕ) such that $2x$ lies in $H(L)$. On the other hand, (cf. Proposition 4 on page 310 of [Mu1,I]), an element x of X lies in $H(L^2)$ if and only if[1] $2x$ lies in $H(L)$. Furthermore,

$$H(L) = 2H(L^2).$$

It follows that an element (x, ψ) of $G(L^4)$ lies in $G(L^4)'_{(2,\alpha)}$ if and only if x lies in $H(L^2)$. If (x, ψ) is such then we can find an isomorphism ϕ of L^2 onto $T_x^* L^2$. The induced isomorphism $\phi^{\otimes 2}$ of L^4 onto $T_x^* L^4$ will differ by a scalar multiple from ψ. By suitably modifying ϕ by a scalar factor, we can suppose that $\phi^{\otimes 2} = \psi$. It follows that $G(L^4)'_{(2,\alpha)}$ is nothing other than the image of $G(L^2)$ under the homomorphism E. The composition $h_{(2,\alpha)} \circ E$ of $h_{(2,\alpha)}$ and E is a homomorphism of $G(L^2)$ onto $G(L)$ which we will denote by η. When it is necessary to indicate the dependence of η on L, we will write η^L instead of η.

  Applying the Mumford functor Mum to the morphisms (Δ, id) and (Δ', γ) of (X, L^4) into $(X \times X, M^2)$ we obtain, among other things, homomorphisms

$$h_{(\Delta, id)} : G(L^4)'_{(\Delta, id)} \to G(M^2)$$

---

[1]   Indeed, for any $x \in X$, the line bundle $T_x^* L$ is isomorphic to $L \otimes \alpha$ for some $\alpha \in Pic^0(X)$. Hence,

$$T_{2x}^* L \simeq T_x^* (L \otimes \alpha) \simeq T_x^* (L) \otimes \alpha \simeq L \otimes \alpha^2.$$

It follows that $2x \in H(L)$ if and only if $\alpha^2 \simeq \mathcal{O}$, i.e. $T_x^* L^2 \simeq L^2$, or, what is the same, $x \in H(L^2)$.

and

$$h_{(\Delta',\gamma)} : G(L^4)'_{(\Delta',\gamma)} \to G(M^2)$$

of $G(L^4)'_{(\Delta,id)}$ and $G(L^4)'_{(\Delta',\gamma)}$ respectively into $G(M^2)$. If x lies in $H(L^2)$ then $\Delta(x) = (x,x)$ and $\Delta'(x) = (x,-x)$ certainly lie in $G(M^2)$. Therefore if γ is as in Lemma (12.6), the homomorphism E maps $G(L^2)$ into $G(L^4)'_{(\Delta,id)}$ and $G(L^4)'_{(\Delta',\gamma)}$. The compositions $h_{(\Delta,id)} \circ E$ and $h_{(\Delta',\gamma)} \circ E$ are therefore defined and will be denoted by अ and $\text{अ}'$ respectively:[2].

$$\text{अ} = h_{(\Delta,id)} \circ E$$
$$\text{अ}' = h_{(\Delta',id)} \circ E$$

Furthermore, both अ and $\text{अ}'$ map $G(L^2)$ into $G(M^2)'_{(\xi,\beta)}$.

Proposition (13.1): *The following diagrams of groups and homomorphisms are commutative.*

(13.2)
$$\begin{array}{ccc}
G(L^2) & \xrightarrow{\eta} & G(L) \\
\downarrow{\scriptstyle \text{अ}} & & \downarrow{\scriptstyle h_{(s_1,id)}} \\
G(M^2)'_{(\xi,\beta)} & \xrightarrow{h_{(\xi,\beta)}} & G(M)
\end{array}$$

(13.3)
$$\begin{array}{ccc}
G(L^2) & \xrightarrow{\eta} & G(L) \\
\downarrow{\scriptstyle \text{अ}'} & & \downarrow{\scriptstyle h_{(s_2,id)}} \\
G(M^2)'_{(\xi,\beta)} & \xrightarrow{h_{(\xi,\beta)}} & G(M)
\end{array}$$

Proof: We are taking α, β and γ as in Lemma (12.6). Applying that lemma, we obtain the commutative diagrams (13.4) and (13.5) given below.

(13.4)
$$\begin{array}{ccc}
G(L^4)'_{(2,\alpha)} & \xrightarrow{h_{(2,\alpha)}} & G(L) \\
\downarrow{\scriptstyle h_{(\Delta,id)}} & & \downarrow{\scriptstyle h_{(s_1,id)}} \\
G(M^2)'_{(\xi,\beta)} & \xrightarrow{h_{(\xi,\beta)}} & G(M)
\end{array}$$

(13.5)
$$\begin{array}{ccc}
G(L^4)_{(2,\alpha)} & \xrightarrow{h_{(2,\alpha)}} & G(L) \\
\downarrow{\scriptstyle h_{(\Delta',\gamma)}} & & \downarrow{\scriptstyle h_{(s_2,id)}} \\
G(M^2)'_{(\xi,\beta)} & \xrightarrow{h_{(\xi,\beta)}} & G(M)
\end{array}$$

Here we have written $G(L)$ in the upper right hand corners of both diagrams since

$$G(L)'_{(s_1,id)} = G(L)'_{(s_2,id)} = G(L).$$

Also we have written $G(M^2)'_{(\xi,\beta)}$ in the lower left hand corners of (13.4) and (13.5) since अ and $\text{अ}'$ both $G(L^4)'_{(2,\alpha)}$ into $G(M^2)'_{(\xi,\beta)}$. On the other hand, by definition

---

[2] The symbol अ is our way of representating the Sanskrit vowel "a" in devanagari script using T$_{\text{E}}$X.

of $3\bar{|}$, $3\bar{|}'$ and η, the following two diagrams are commutative.

(13.6)

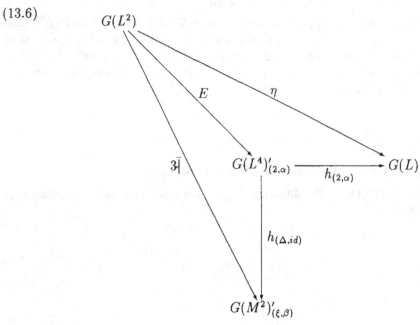

(13.7)

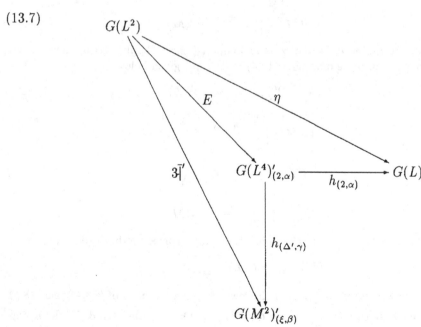

If we now splice (13.6) onto (13.4) and (13.7) onto (13.5) we obtain the commutative diagrams (13.8) and (13.9) below. From these the proposition follows at once by looking at the morphisms which form the perimeters of (13.8) and (13.9).

(13.8)

(13.9)

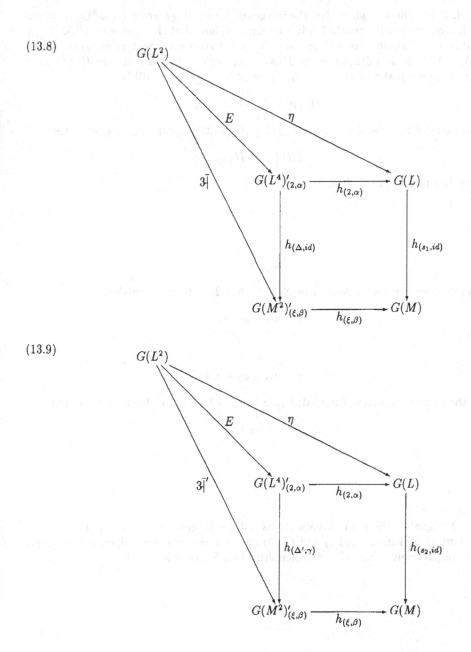

Proposition (13.10):   *The commutative diagrams (13.2) and (13.3) characterize the homomorphism*

$$h_{(\xi,\beta)} : G(M^2)'_{(\xi,\beta)} \to G(M).$$

Proof: It is enough to show that the images of $3\bar{|}$ and $3\bar{|}'$ generate $G(M^2)_{(\xi,\beta)}$. Since the images obviously contain $k^\times$, it is enough to show that they generate $G(M^2)_{(\xi,\beta)}$ modulo $k^\times$. Therefore we will be done if we can show that whenever (x,y) is a point of $X \times X$ such that $\xi(x,y)$ lies in $H(M)$ then we can find u and v in $H(L^2)$ such that (x,y) is equal to $\Delta(u) + \Delta'(v)$. Since $M = L \boxtimes L$, we will have

$$\xi(x,y) = (x+y, x-y)$$

if and only if $x+y$ and $x-y$ lie in $H(L)$. By [Mu1,I], prop.4, p.310, we have

$$2H(L^2) = H(L),$$

so we can find u_0, v_0 in $H(L^2)$ such that

$$2u_0 = x+y$$

and

$$2v_0 = x-y$$

Therefore we can find a point z in X such that $2z = 0$ and such that

$$x = u_0 + v_0 + z$$

and

$$y = u_0 - v_0 + z.$$

By the same proposition from [Mu1,I], z belongs to $H(L^2)$. Therefore, writing

$$u = u_0 + z$$

and

$$v = v_0,$$

we are done.

Our goal in the next three sections will be to construct isomorphisms of the commutative diagrams (13.2) and (13.3) onto the commutative diagrams (8.3) and (8.4) respectively. This will be accomplished by Proposition 17.9.

§14   Characterization of $\overline{\mathfrak{A}}$, $\overline{\mathfrak{A}}'$, $\mathbf{h}_{(s_1,\mathrm{id})}$ and $\mathbf{h}_{(s_2,\mathrm{id})}$

We will retain the assumptions and notation of the preceding section. In this section, we will resolve $\overline{\mathfrak{A}}$, $\overline{\mathfrak{A}}'$, $h_{(s_1,id)}$ and $h_{(s_2,id)}$ into simpler elements. First we define the homomorphism

$$\mu_1 : G(L) \times G(L) \to G(M)$$

of $G(L) \times G(L)$ into $G(M)$ by the rule

(14.1) $$\mu_1\left((x,\phi),(y,\psi)\right) = \left((x,y),\phi \boxtimes \psi\right)$$

where $\phi \boxtimes \psi$ denotes the isomorphism

$$p_1^*(\phi) \otimes p_2^*(\psi) : M = p_1^*(L) \otimes p_2^*(L) \to T_{(x,y)}^*(M) = p_1^*(T_x^*(L)) \otimes p_2^*(T_y^*(L))$$

of $M = p_1^*(L) \otimes p_2^*(L)$ onto $T_{(x,y)}^*(M) = p_1^*(T_x^*(L)) \otimes p_2^*(T_y^*(L))$. Similarly, we define the homomorphism

$$\mu_2 : G(L^2) \times G(L^2) \to G(M^2)$$

of $G(L^2) \times G(L^2)$ onto $G(M^2)$ by

(14.2) $$\mu_2\left((x,\phi),(y,\psi)\right) = \left((x,y),\phi \boxtimes \psi\right)$$

whenever (x,ϕ) and (y,ψ) both belong to $G(L^2)$. It is very easy to see that μ_1 and μ_2 are both surjective and that the kernel of each consists of all (s,t) in $k^\times \times k^\times$ such that $st = 1$. Denote by

$$\Delta_G : G(L^2) \to G(L^2) \times G(L^2)$$

the diagonal morphism of $G(L^2)$ into $G(L^2) \times G(L^2)$. Then we have

(14.3) $$\mu_2 \circ \Delta_G = \overline{\mathfrak{A}}.$$

Since L is symmetric, we have that L and $\iota^*(L)$ are isomorphic. Let

$$\psi : L \to \iota^*(L)$$

be the normalized isomorphism of L onto $\iota^*(L)$ as explained in the remarks following Definition 11.3. Then for every (x,ϕ) in $G(L)$ the composition

$$(T_{-x}^*\psi)^{-1} \circ (\iota^*\phi) \circ \psi : L \to T_{-x}^*(L)$$

is an isomorphism. Furthermore, the mapping

$$\omega_L : G(L) \to G(L)$$

of $G(L)$ onto itself given by

(14.4) $$\omega_L(x,\phi) = \left(-x, (T_{-x}^*\psi)^{-1} \circ (\iota^*\phi) \circ \psi\right)$$

is an automorphism of $G(L)$. We will simplify notation by writing ω instead of ω_L.

Since L^2 is also symmetric, the same considerations apply to L^2. In particular, we obtain an automorphism ω' of $G(L^2)$ by the rule

(14.5) $$\omega'\left((x, \phi)\right) = (-x, (T^*_{-x}\psi^{\otimes 2}) \circ (\iota^*\phi) \circ \psi^{\otimes 2})$$

for (x, ϕ) in $G(L^2)$. Here we have written ω' instead of the more complicated notation ω_{L^2}.

Define the homomorphism

$$\Delta'_G : G(L^2) \to G(L^2) \times G(L^2)$$

of $G(L^2)$ into $G(L^2) \times G(L^2)$ by the rule

(14.6) $$\Delta'_G = (1_{G(L^2)} \times \omega') \circ \Delta_G.$$

Then we have

(14.7) $$\mu_2 \circ \Delta'_G = 3_|^{\vec{\jmath}'}.$$

Denote by i_1 and i_2 the homomorphisms

$$i_1, i_2 : G(L) \to G(L) \times G(L)$$

of $G(L)$ into $G(L) \times G(L)$ given by

(14.8) $$i_1 = (z, 1)$$
$$i_2 = (1, z).$$

Then we have

(14.9) $$\mu_1 \circ i_1 = h_{(s_1, id)}$$
$$\mu_1 \circ i_2 = h_{(s_2, id)}$$

By Proposition 13.10, the homomorphism $h_{(\xi, \beta)}$ will be known as soon as the homomorphisms $3_|^{\vec{\jmath}}$, $3_|^{\vec{\jmath}'}$, $h_{(s_1, id)}$, $h_{(s_2, id)}$ and η are known. Using equations (14.3), (14.7) and (14.9), it follows that $h_{(\xi, \beta)}$ will be understood as soon as the homomorphisms Δ_G, Δ'_G, μ_1, μ_2, i_1, i_2 and η are understood.

Chapter III

Theta Structures and the Addition Formula

§15 Symmetric Theta Structures

If $\delta = (d_1, \ldots, d_g)$ is any element of S, we define the mapping

$$D_\delta : G(\delta) \to G(\delta)$$

from $G(\delta)$ onto itself by the rule

(15.1) $$D_\delta(\alpha, x, l) = (\alpha, -x, -l).$$

It is easy to verify that D_δ is an automorphism of order 2 of $G(\delta)$. When it is not necessary to specify δ, we will write D instead of D_δ.

Let Z be an abelian variety of dimension n and let N be an ample line bundle on Z. Then as we have already remarked in §9 following Diagram 9.2, there is an element δ of S and an isomorphism of $G(N)$ onto $G(\delta)$ which is the identity on $k^\times$. Such an isomorphism is called a **theta structure** on N. Suppose that N is symmetric. Then there is an isomorphism ψ of N onto $\iota^*(N)$. If we fix ψ, then according to (14.4) we can determine an automorphism ω_N of $G(N)$ be the rule

(15.2) $$\omega_N(x, \phi) = \left(-x, \, (T^*_{-x}\psi)^{-1} \circ (\iota^* \phi) \circ \psi \right).$$

We will say that a theta structure $f : G(N) \to G(\delta)$ is **symmetric** if

(15.3) $$f \circ \omega_N = D \circ f.$$

Each of the groups $G(N)$, $G(\delta)$ is a group whose center is $k^\times$ and which is a finite abelian group modulo $k^\times$. We will say that a group with these properties is a **group of Heisenberg type**. If G is a group of Heisenberg type and if $H = G/k^\times$, we will say that **H is the factor group of G**. There is a natural pairing

$$B : H \times H \to k^\times$$

of $H \times H$ into $k^\times$, that is to say, a bicharacter of H, defined by

(15.4) $$B(x, y) = \tilde{x} \cdot \tilde{y} \cdot \tilde{x}^{-1} \cdot \tilde{y}^{-1}$$

where $\tilde{x}$ and $\tilde{y}$ are any representatives of x and y in G. When it is important to specify G, we will write B_G instead of B. It is straightforward to check that B is a well-defined, nondegenerate and alternating[3] pairing of $H \times H$ into $k^\times$. We call B the **natural bicharacter** of H. If H_1 and H_2 are finite abelian groups and if B_1 and B_2 are nondegenerate alternating bicharacters of H_1 and H_2 respectively, then a homomorphism

$$\sigma : H_1 \to H_2$$

from H_1 to H_2 will be called **symplectic** with respect to H_1 and H_2 if

(15.5) $$B_1(x, y) = B_2 \left(\sigma(x), \sigma(y) \right)$$

for all x and y in H_1. Suppose that G_1 and G_2 are groups of Heisenberg type and $g : G_1 \to G_2$ is an isomorphism which is the identity on $k^\times$. Taking H_i to be the factor group of G_i and B_i to be the natural bicharacter of H_i for $i = 1, 2$, we have that g induces a symplectic isomorphism of H_1 onto H_2. In particular, any theta structure

$$f : G(N) \to G(\delta)$$

induces a symplectic isomorphism of $H(N)$ onto $H(\delta)$.

Proposition (15.6):   *Let (V, N) be an object of the category $\mathcal{A}$ of §10. Then every symplectic isomorphism of $(H(N), B_{G(N)})$ onto $(H(\delta), B_{G(\delta)})$ is induced by a theta structure*

$$f : G(N) \to G(\delta)$$

of N. If N is strongly symmetric, we may require f to be a symmetric theta structure.

Proof: The first assertion is proved in [Mu1, I],p.318 (see the first paragraph of the proof of Remark 2). To prove the second assertion, we will show that the rest of Mumford's proof of Remark 2 goes through in case N is totally symmetric. In fact, let σ be a symplectic isomorphism of $(H(N), B_{G(N)})$ onto $(H(\delta), B_{G(\delta)})$ and let

$$g : G(N) \to G(\delta)$$

be a theta structure which induces

$$\sigma : H(N) \to H(\delta).$$

Let D' denote the automorphism

$$g \circ \omega \circ g^{-1} : G(\delta) \to G(\delta)$$

of $G(\delta)$. Then D' is the identity on $k^\times$ and induces $-1_{H(\delta)}$ on $H(\delta)$. Therefore, since D has the same property, it follows that $D'D^{-1}$ is an automorphism of $G(\delta)$ inducing the identity on $k^\times$ and on $H(\delta)$. We must have

(15.7) $$D' \circ D^{-1}(\alpha, x, l) = (\alpha \cdot \chi(x, l), x, l)$$

for all (α, x, l) in $G(\delta)$, where χ is a character of $H(\delta)$. By Proposition 3 on p.309 of [Mu1,I], whenever z is an element of $G(N)$ whose image w in $H(N)$ satisfies $2w = 0$,

---

[3] in the sense that $B(x, x) = 1$ for all $x \in H$

we have

$$\omega_N(z) = \epsilon_N(w) \cdot z,$$

where ϵ_N is the quadratic character of $H(N)$ which we defined in §11. Since N is strongly symmetric,

$$\epsilon_N(w) = 1$$

for all such w and we have $\omega_N(z) = z$ for all such z. It follows that if $z' = (\alpha, x, l)$ is an element of $G(\delta)$ such that $2x = 0$ and $l^2 = 1$ then

$$D'(z') = z'.$$

But for such z' we also have

$$D(z') = z',$$

so that

$$z' = D' \circ D^{-1}(z') = (\alpha \cdot \chi(x, l), x, l).$$

It follows that the character χ is trivial on the subgroup of $H(\delta)$ consisting of elements u such that $2u = 0$. Consequently, we can write $\chi = \psi^2$ where ψ is a character of $H(\delta)$. Denote by F_ψ the automorphism of $G(\delta)$ given by

$$F_\psi(\alpha, x, l) = (\alpha\psi(x, l)^{-1}, x, l).$$

Then

(15.8) $$F_\psi \circ D' \circ F_\psi^{-1}(\alpha, x, l) = \left(\alpha \cdot \chi(x, l) \cdot \psi^{-2}(x, l), \, -x, -l \right) = D(\alpha, x, l).$$

If we now let $f = F_\psi \circ g$, we have

(15.9) $$f \circ \omega_N = D \circ f$$

which says that f is a symmetric theta structure. Since g induces

$$\sigma : H(N) \to H(\delta)$$

and F_ψ induces $1_{H(\delta)}$ on $H(\delta)$, it follows that f induces

$$\sigma : H(N) \to H(\delta)$$

and we are done.

§16   The Arf Invariant and the Existence of Symmetric Theta Structures

Let (X, L) be a strongly symmetric object of the category $\mathcal{A}$. Then by definition we have

$$\epsilon_L(x) = 1$$

for all x in $H(L)$ such that $2x = 0$. The bicharacter associated to ϵ_L is denoted by $\beta(x, y)$, as in §11. Then β is a bicharacter of X_2. On the other hand, $H(L^2)$ is the factor group of $G(L^2)$ in the sense of §15, so it has a natural alternating form

$B_{G(M)}$, where $M = L^2$. According to Corollary 1 of Proposition 6 on pages 314-315 of [Mu1,I], the bicharacter β is the restriction of $B_{G(M)}$ to X_2.

Lemma (16.1):   Let H be a finite abelian group. Let

$$b : H \times H \to k^\times$$

be a nondegenerate bicharacter and let b' denote the restriction of b to $H_2 \times H_2$, where H_2 denotes the subgroup of H consisting of all h such that $2h = 0$. Then the radical of b' is $H_2 \cap 2 \cdot H$.

Proof: Let us write J to denote the group $H_2 \cap 2 \cdot H$. If a is an element of J, we can write $a = 2 \cdot a'$ with a' in H. Therefore for all a'' in H_2 we have

$$(16.2) \qquad b'(a, a'') = b'(2 \cdot a', a'') = b'(a', 2 \cdot a'') = b'(a', 0) = 1.$$

This proves that J is contained in the radical. Conversely, supposed that a belongs to H_2 but not to $2 \cdot H$. Then the character χ_a of H given by

$$\chi_a(x) = b(a, x)$$

does not lie in $2 \cdot H^*$, and therefore χ_a represents a nontrivial element of

$$H^*/2 \cdot H^* = (H_2)^*.$$

Therefore we can find an element a' of H_2 such that

$$b'(a, a') = \chi_a(a') \neq 1.$$

So a does not lie in the radical of b'.

Corollary (16.3):   Let L be a strongly symmetric line bundle on an abelian variety X and let β be the bicharacter of X_2 defined in §11. Then the radical of β is $H(L) \cap X_2$. In particular, if $H(L)$ has odd order then β is a nondegenerate bicharacter of X_2.

Proof: The second assertion is a trivial consequence of the first. On the other hand, let $H = H(M)$, where $M = L^2$. Then $H_2 = X_2$. Furthermore, since L^2 is ample, the Weil form $B_{G(M)}$ is nondegenerate. Therefore, if we take b to be $B_{G(M)}$ in Lemma 16.1, we see that the radical of β is $X_2 \cap 2H(M)$. But by Proposition 4 on p.310 of [Mu1,I], we have

$$2 \cdot H(M) = H(L),$$

so we are done.

Corollary (16.4):   We retain the notation of Corollary 16.3. The quadratic character ϵ_L induces a nondegenerate quadratic character of $X_2/(H(L) \cap X_2)$.

Proof: This is a trivial consequence of Corollary 16.3 and of the fact that β is the bilinear form associated to ϵ_L.

We may regard X_2 as a vector space over the field $\mathbf{F}_2$ with two elements. Since ϵ_L can take on only the values ± 1, the quadratic character ϵ_L may be written in the form $(-1)^{Q_L}$, where Q_L is a quadratic form on the $\mathbf{F}_2$ vector space X_2. By

Corollary 16.4, Q_L induces a nondegenerate quadratic form, denoted Q'_L on the $\mathbf{F_2}$ vector space $X_2/(H(L)\cap X_2)$. Since the bicharacter β is alternating, Corollary 16.4 implies that the bilinear form associated to Q'_L is a nondegenerate alternating form. Therefore we can find a basis

$$e_1,\ldots,e_{2g}$$

for $X_2/(H(L)\cap X_2)$ with respect to which Q'_L has the form

(16.5)
$$Q'_L\left(\sum_{i=1}^{2g} x_i\, e_i\right) = \sum_{i=1}^{g} x_i\, x_{i+g} + \sum_{i=1}^{2g} a_i\, x_i.$$

The element

$$\sum_{i=1}^{g} a_i a_{g+i}$$

of $\mathbf{F}_2$ depends only on the quadratic form Q'_L and is called the **Arf invariant of the quadratic form $\mathbf{Q_L}$**. For our purposes, it is convenient to rephrase this in terms of ϵ_L and β. Let K_1 and K_2 be maximal isotropic subgroups of $H(L^2)$ with respect to β such that

$$K_1 \cap K_2 = (0)$$

and

$$K_1 + K_2 = H(L^2).$$

For x in X_2, write

$$x = x_1 + x_2$$

with x_i in K_i for $i = 1,2$ and define

$$\epsilon'_L(x) = \beta(x_1, x_2).$$

Then ϵ'_L is a quadratic character of $H(L)$ valued in $\{\pm 1\}$ and its associated bicharacter is β. We can therefore write

$$\epsilon_L(x) = \epsilon'_L(x) \cdot \chi(x)$$

where χ is a character of X_2 trivial on $H(L)\cap X_2$. The character χ can be written in the form

$$\chi(x) = \beta(x, y)$$

for suitable y in $H(L)$. If we write

$$y = y_1 + y_2$$

with y_i in K_i for $i = 1,2$, then we define the **Arf invariant of the quadratic character ϵ_L** to be $\beta(y_1, y_2)$. It is easy to see that the Arf invariant of of the quadratic character ϵ_L is $(-1)^a$ where a is the Arf invariant of the quadratic form Q_L. The Arf invariant of the quadratic character ϵ_L is called the **Arf invariant of L** and is denoted $Arf(L)$. Note that $Arf(L)$ takes on the values ± 1, whereas the Arf invariant of the quadratic form Q_L takes on the values 0 and 1. If (V_1, Q_1) and (V_2, Q_2) are quadratic vector spaces over $\mathbf{F}_2$ of the same dimension, they are isomorphic if and only if they have the same Arf invariant. One may be tempted to identify the additive group of $\mathbf{F}_2$ with the multiplicative group $\{\pm 1\}$ and regard the Arf invariants of quadratic characters of $\mathbf{F}_2$ vector spaces as being the same thing as Arf invariants of quadratic forms of $\mathbf{F}_2$ vector spaces. Natural as it may seem, to do so causes a problem: the expression "Arf invariant 1" has one meaning for a quadratic character and exactly the opposite meaning for a quadratic form. We will attempt to avoid confusion by carefully distinguishing between quadratic forms and quadratic characters. We will say that **L has trivial Arf invariant** if $Arf(L) = 1$, or what is the same, if the Arf invariant of Q_L is zero.

Lemma (16.6): *Let L be a strongly symmetric line bundle on an abelian variety X let ϵ' be the quadratic character which ϵ_L induces on $X_2/(H(L) \cap X_2)$. Let Q be a quadratic character on $H(2\delta)_2/H(\delta)_2$ and suppose that the Arf invariant of Q equals the Arf invariant of the quadratic character ϵ'. Then there is a symmetric theta structure*

$$f : G(L^2) \to G(2\delta)$$

which induces a quadratic isomorphism of

$$(X_2/(H(L) \cap X_2),\ \epsilon') \to (H(2\delta)_2/H(\delta)_2,\ Q).$$

Proof: We may as well suppose that $\delta = (d_1, \ldots, d_g)$ where g is the dimension of X and that modulo 2 we have

$$(16.7) \qquad\qquad d_i \equiv \begin{cases} 1 \ (\text{mod } 2) & \text{if } 1 \le i \le k \\[2mm] 0 \ (\text{mod } 2) & \text{if } k+1 \le i \le g \end{cases}$$

for some k such that $0 \le k \le g$. Let $f' : G(L^2) \to G(2\delta)$ be any symmetric theta structure. Then f' induces a symplectic isomorphism

$$\sigma' : H(L^2) \to H(2\delta)$$

of $H(L^2)$ onto $H(2\delta)$. Let R denote the radical of β and let R' denote the radical of the restriction of $B_{G(2\delta)}$ to $H(2\delta)_2$. Then we have

$$(16.8) \qquad \begin{aligned} R &= H(L) \cap X_2 &&= H(L) \cap H(L^2)_2 \\[2mm] R' &= H(2\delta)_2 \cap H(\delta) &&= H(\delta)_2 \end{aligned}$$

Let

$$U = X_2/R$$

and

$$U' = H(2\delta)_2/R'.$$

Then σ' induces a symplectic isomorphism of U onto U'. Using this symplectic isomorphism, we can carry ϵ' over to a quadratic character on U' which we will denote by Q'. Clearly, Q and Q' have the same Arf invariant. Consequently, there is an automorphism σ of U' which transforms Q' into Q. Since σ must preserve the associated nondegenerate alternating forms, which happen to coincide in this case since σ' is symplectic, it follows that σ is symplectic. If we can lift σ to a symplectic automorphism σ'' of $H(2\delta)$ then $\sigma'' \circ \sigma'$ will lift, by Proposition 15.6, to a symmetric theta structure $f : G(L^2) \to G(2\delta)$ satisfying the conditions of the lemma.

Write $H(\delta) = H(\delta') \oplus H(\delta'')$ where

(16.9)
$$\delta' = (d'_1, \ldots, d'_g)$$
$$\delta'' = (d''_1, \ldots, d''_g)$$

and where for $1 \leq i \leq g$ we define d'_i to be the largest power of 2 dividing d_i and d''_i to be the largest odd factor of d_i. Then the inclusion of $H(2\delta')$ into $H(2\delta)$ induces an isomorphism of $H(2\delta')_2/H(\delta')_2$ onto $H(2\delta)_2/H(\delta)_2$. But we have

(16.10)
$$H(2\delta') = K(2\delta') \oplus K(2\delta')^*$$

and

(16.11)
$$K(2\delta') = \overset{g}{\underset{i=1}{\oplus}} K(2^{e_i})$$

where e_i is 0 for $1 \leq i \leq k$ and $e_i > 0$ for $k+1 \leq i \leq g$. If all e_i are greater than zero then $H(2\delta')_2/H(\delta')_2 = 0$ and there is nothing to prove. Therefore we may suppose that $k > 0$. Let

(16.12)
$$A = \overset{k}{\underset{i=1}{\oplus}} K(2^{1+e_i}) = \overset{k}{\underset{i=1}{\oplus}} K(2)$$

and

(16.13)
$$B = \overset{g}{\underset{i=k+1}{\oplus}} K(2^{1+e_i})$$

Then $K(2\delta') = A \oplus B$ and the inclusion of $A \oplus A^*$ into $H(2\delta')$ induces an isomorphism of $A \oplus A^*$ onto $H(2\delta')_2/H(\delta')_2$. Since $H(2\delta')$ is the direct sum of $A \oplus A^*$ and $B \oplus B^*$, we may extend the symplectic automorphism σ to a symplectic automorphism σ'' of $H(2\delta)$ by taking σ'' to be σ on $A \oplus A^*$ and the identity mapping on $B \oplus B^* \oplus H(\delta'')$.

Lemma (16.14): Let G be a group of Heisenberg type with center $k^\times$ and factor group H and let $B = B_G$ denote the associated bicharacter of H. Then $H_2 \cap 2H$ is the radical of the restriction B' of B to H_2. Let B' also denote the bicharacter of $H_2/(H_2 \cap 2H)$ induced by B. Let us write $\bar{z}$ to denote the image in H of an element z of G. Let Q be a quadratic character on $H_2/(H_2 \cap 2H)$ whose associated bicharacter is B'. Let G_Q denote the subset of G consisting of all z in G such that z^2 lies in $k^\times$ and equals $Q(\bar{z})$. Then G_Q is a normal subgroup of G and G/G_Q is a group of Heisenberg type with center $k^\times$ and factor group $2H$. Furthermore,

$$G_Q \cap k^\times = \{\pm 1\}.$$

Proof: Since G is of Heisenberg type, the associated bicharacter B is nondegenerate. Therefore, by Lemma 16.1, the radical of B' is $H_2 \cap 2H$. Suppose that x belongs to G_Q and that y is an element of G. Then we have

$$2\overline{x} \quad = \quad 0$$

(16.15)
$$x^2 \quad = \quad Q(\overline{x})$$

$$(yxy^{-1})^2 \quad = \quad yx^2y^{-1} \quad = \quad Q(\overline{x})$$

since $k^\times$ is the center of G. This shows that G_Q is invariant under inner automorphisms of G. To see that G_Q is a subgroup of G, suppose that x and y both lie in G_Q. Then we have

$$
\begin{aligned}
(xy)^2 \quad &= \quad xyxy \\
&= \quad xyx^{-1}x^2y^2y^{-1} \\
&= \quad Q(\overline{x})Q(\overline{y})xyx^{-1}y^{-1} \\
&= \quad Q(\overline{x})Q(\overline{y})B'(\overline{x},\overline{y}) \\
&= \quad Q(\overline{x})Q(\overline{y})Q(\overline{x}+\overline{y})Q(\overline{x})^{-1}Q(\overline{y})^{-1} \\
&= \quad Q(\overline{x}+\overline{y})
\end{aligned}
$$

which shows that G_Q is closed under products in G. It is easy to see that $G_Q \cap k^\times = \{\pm 1\}$. Therefore, since H has finite order and G_Q is closed under products, every element of G_Q has finite order. It follows that G_Q is closed under inverses, so G_Q is a subgroup.

Let G' denote the group G/G_Q. The center $k^\times$ of G maps into the center of G'. Denote by T the image of $k^\times$ in G'. Then since $k^\times \cap G_Q = \{\pm 1\}$, the group T is isomorphic to $k^\times/\{\pm 1\}$ and therefore to $k^\times$ since $\{\pm 1\}$ is the kernel of the endomorphism $z \mapsto z^2$ of $k^\times$. Hence we can identify T with $k^\times$ in a natural way. Since $G/k^\times$ is abelian, so is G'/T. Therefore we will know that G' is of Heisenberg type as soon as we know that the center of G' is precisely T. Let x be an element of G and suppose that for all y in G the commutator $xyx^{-1}y^{-1}$ lies in G_Q. Then we have that

$$B(\overline{x},\overline{y}) = xyx^{-1}y^{-1}$$

lies in

$$G_Q \cap k^\times = \{\pm 1\}$$

for all y in G. Therefore,

$$B(2\overline{x},\overline{y}) = 1$$

for all $\overline{y}$ in H. Since B is nondegenerate, we therefore have $2\overline{x} = 0$ which implies that x^2 belongs to $k^\times$. Choose an element t of $k^\times$ such that

$$(t^{-1}x)^2 = Q(\overline{x}).$$

Then $t^{-1}x$ belongs to G_Q and therefore the coset xG_Q belongs to T. This proves that T is the center of G' and that G' is of Heisenberg type. In the same way, we see that the image of G_Q in H is all of H_2: if z is an element of H such that $2z = 0$, choose w in G such that $\overline{w} = z$. Then for suitable t in $k^\times$, the element tw lies in G_Q.

To compute the factor group H' of G', note that

$$(16.16) \qquad H' = G'/T \cong G/k^\times G_Q = H/H_2 \cong 2H$$

and the induced mapping of H onto H' can be identified with the mapping of H onto $2H$ given by $x \mapsto 2x$. This completes the proof of Lemma 16.14.

Corollary (16.17): *Let L be an ample symmetric invertible sheaf of separable type on the abelian variety X and let Q denote the quadratic character ϵ_L of X_2. Let $G = G(L^2)$ and let G_Q be as in Lemma 16.14. Then η maps G onto $G(L)$ with kernel G_Q.*

Proof: We know that η is surjective since $2H(L^2) = H(L)$, so we just have to prove that the kernel of η is G_Q. Both G_Q and the kernel of η have the same image in $H(L^2)$, namely X_2. Let w be an element of G whose image in $H(L^2)$ is a point x of order 2. Then we can find an element t of $k^\times$ such that $z = t^{-1}w$ is an element of order 2 in G. Applying Proposition 6 on p.312 of [Mul,I], we have

$$(16.18) \qquad \eta(w) = t^2\eta(z) = \lambda^2\epsilon_L(x) = w^2 Q(x).$$

Therefore $\eta(x) = 1$ if and only if $w^2 = Q(x)$.

Proposition (16.19): *Let L be an ample strongly symmetric invertible sheaf of separable type on the abelian variety X. Let Q denote the quadratic character induced on $X_2/(H(L) \cap X_2)$ by ϵ_L. Let (x, χ) be an element of $H(2\delta)$ such that $2x = 0$ and $\chi^2 = 1$ and assume that $\chi(x)$ is equal to the Arf invariant $Arf(L)$ of the quadratic character ϵ_L. Let*

$$f_1 : G(L) \to G(\delta)$$

be a symmetric theta structure on L. Then there exists a symmetric theta structure

$$f_2 : G(L^2) \to G(2\delta)$$

on L^2 such that the diagram

$$(16.20) \qquad \begin{array}{ccc} G(L^2) & \xrightarrow{f_2} & G(2\delta) \\ \downarrow{\scriptstyle \eta} & & \downarrow{\scriptstyle \eta_{(x,\chi)}} \\ G(L) & \xrightarrow{f_1} & G(\delta) \end{array}$$

is commutative, where $\eta_{(x,\chi)}$ is the homomorphism defined by equation (4.1).

Proof: For $i = 1, 2$ we denote by G_i the group of all symmetric automorphisms of $G(i\delta)$ inducing the identity automorphism on $k^\times$. Denote by Γ_1 the set of all symmetric theta structures on L and by Γ_2 the set of all symmetric theta structures on L^2. Then for $i = 1, 2$, Γ_i is a principal homogeneous space for the group G_i. Let

Γ denote the set of all commutative diagrams

(16.21)
$$
\begin{array}{ccc}
G(L^2) & \xrightarrow{f_2'} & G(2\delta) \\
\downarrow{\scriptstyle\eta} & & \downarrow{\scriptstyle\eta_{(x,\chi)}} \\
G(L) & \xrightarrow{f_1'} & G(\delta)
\end{array}
$$

where f_1' and f_2' are symmetric theta structures. There are natural projections of Γ onto Γ_i for $i = 1, 2$ and our task is to prove that the projection of Γ onto Γ_1 is surjective. We will first prove that Γ is nonempty. Denote by Q' the quadratic character on $H(2\delta)_2/H(\delta)_2$ given by

(16.22) $$Q'(y, m) = m(y) \cdot \chi(y) \cdot m(x).$$

Then Q' has Arf invariant $\chi(x)$, which by hypothesis equals the Arf invariant of Q. By Lemma 16.6 there is a symmetric theta structure

$$f_2' : G(L^2) \to G(2\delta)$$

which induces a quadratic isomorphism

$$g : X_2/(H(L) \cap X_2) \to H(2\delta)_2/H(\delta)_2$$

of $X_2/(H(L) \cap X_2)$ onto $H(2\delta)_2/H(\delta)_2$ carrying Q onto Q'. By Corollary 16.17, the group $G(L^2)_Q$ is the kernel of η. On the other hand, by Lemma 4.2, $G(2\delta)_{Q'}$ is the kernel of $\eta_{(x,\chi)}$. Therefore f_2' induces a theta structure

$$f_1' : G(L) \to G(\delta)$$

such that

(16.23) $$f_1' \circ \eta = \eta_{(x,\chi)} \circ f_2'.$$

Then f_1' is symmetric. In fact, if w is an element of $G(L)$, write $w = \eta(z)$ with z in $G(L^2)$. Then we have, writing M for L^2,

(16.24)
$$
\begin{aligned}
D_\delta \circ f_1'(w) &= D_\delta \circ f_1' \circ \eta(z) \\
&= D_\delta \circ \eta_{(x,\chi)} \circ f_2'(z) \\
&= \eta_{(x,\chi)} \circ D_{2\delta} \circ f_2'(z) \\
&= \eta_{(x,\chi)} \circ f_2' \circ \omega_M(z) \\
&= f_1' \circ \eta \circ \omega_M(z) \\
&= f_1' \circ \omega_L \circ \eta(z) \\
&= f_1' \circ \omega_L(w)
\end{aligned}
$$

Here the identity $\eta \circ \omega_M = \omega_L \circ \eta$ follows from Proposition 5 on p.311 of [Mu1,I].

This proves that the set Γ is nonempty. Next, denote by $\mathcal{G}$ the group whose elements are commutative diagrams

(16.25)
$$
\begin{array}{ccc}
G(2\delta) & \xrightarrow{\sigma_2} & G(2\delta) \\
\downarrow{\scriptstyle \eta_{(x,\chi)}} & & \downarrow{\scriptstyle \eta_{(x,\chi)}} \\
G(\delta) & \xrightarrow{\sigma_1} & G(\delta)
\end{array}
$$

where σ_1 and σ_2 are symmetric automorphisms inducing the identity mapping on $k^\times$. We can identify such a diagram with the pair (σ_1, σ_2) and note that as such $\mathcal{G}$ is a subgroup of the product $G_1 \times G_2$ of the groups G_1 and G_2. It is then easy to see that Γ is a principal homogeneous space for $\mathcal{G}$. We will therefore be done if we can prove that the projection of $\mathcal{G}$ onto G_1 is surjective.

For $i = 1, 2$ and denote by Sp_i the group of symplectic automorphisms of $H(i\delta)$ with respect to the natural symplectic structure on $H(i\delta)$ (cf. [W1], §3). Every symplectic automorphism of $H(i\delta)$ is induced by a symmetric automorphism of $G(i\delta)$ and the natural homomorphism $Sp(2\delta) \to Sp(\delta)$ is surjective. Therefore, we will be done if we can prove that every symmetric automorphism of $G(\delta)$ inducing the identity automorphism of $H(\delta)$ lifts to an element of $\mathcal{G}$. Such a symmetric automorphism must be of the form

(16.26) $$F_\sigma(\alpha, x, l) = (\alpha \cdot \sigma(x, l), x, l)$$

where σ is a character of $H(\delta)$ such that $\sigma^2 = 1$. It is evidently enough to prove this in the case where δ consists of only one number, say, $\delta = (d)$. If d is odd, then we have $\sigma = 1$ and therefore F_σ is the identity automorphism, which obviously lifts. Suppose d is even and let ζ_d and ζ_{2d} be primitive d-th and $2d$-th roots of unity respectively such that

(16.27) $$\zeta_{2d}^2 = \zeta_d.$$

For $i = 1, 2$, we have an isomorphism of $K(id)$ onto $K(id)^*$ given by $v \mapsto \chi_v^{(i)}$ where for all w in $K(d)$ we have

(16.28) $$\chi_v^{(i)}(w) = \zeta_{id}^{v \cdot w}.$$

We can then write σ in the form

$$\lambda(u, \chi_v^{(1)}) = \zeta_d^{a \cdot u + b \cdot v}$$

where a and b are integers such that $2a$ and $2b$ are both divisible by d. Let γ be the endomorphism of $H(2d)$ given by

$$\gamma(r, \chi_s^{(2)}) = (r + 2bs, \chi_{s+2ar}^{(2)}).$$

Then γ is a symplectic automorphism of $H(2d)$. Let $\tilde{\gamma}$ be a lifting of γ to a symmetric automorphism of $G(2d)$. Then the pair $(\tilde{\gamma}, \sigma)$ is an element of $\mathcal{G}$ mapping onto σ, which proves the proposition.

§17   The Addition Formula and the Duplication Formula

Let X be an abelian variety and let L be an ample strongly symmetric invertible sheaf of separable type on X. Let

$$f_1 : G(L) \to G(\delta)$$

be a symmetric theta structure on L. Let (x, χ) be an element of order 2 of $H(2\delta)$ such that $\chi(x)$ is equal to the Arf invariant of L. Then by Proposition 16.19 we can find a symmetric theta structure

$$f_2 : G(L^2) \to G(2\delta)$$

such that the diagram

$$(17.1) \qquad \begin{array}{ccc} G(L^2) & \xrightarrow{f_2} & G(2\delta) \\ \downarrow{\scriptstyle \eta} & & \downarrow{\scriptstyle \eta_{(x,\chi)}} \\ G(L) & \xrightarrow{f_1} & G(\delta) \end{array}$$

is commutative. Using f_1 and f_2 we can transfer to $G(\delta)$ and $G(2\delta)$ and certain related groups the homomorphisms considered in §13 and §14. For example, it is very easy to see that the inclusions i_1 and i_2 of $G(L)$ into the first and second factors respectively of $G(L) \times G(L)$ are transformed by f_1 into the inclusions of $G(\delta)$ into the first and second factors respectively of $G(\delta) \times G(\delta)$. Similarly, Δ_G and Δ'_G are respectively transformed by f_2 into the the homomorphisms $\Delta_{2\delta}$ and $\Delta'_{2\delta}$ defined by

$$(17.2) \qquad \begin{array}{rcl} \Delta_{2\delta}(z) & = & (z, z) \\[2mm] \Delta'_{2\delta}(z) & = & (z, D_{2\delta}(z)). \end{array}$$

For $i = 1, 2$ denote by

$$\mu_{i\delta} : G(i\delta) \times G(i\delta) \to G(i\delta, i\delta)$$

the homomorphism of $G(i\delta) \times G(i\delta)$ into $G(i\delta, i\delta)$ given by

$$(17.3) \qquad \mu_{i\delta}\left((\alpha, y, m), (\beta, z, n)\right) = (\alpha\beta, y, z, m, n).$$

The kernel of $\mu_{i\delta}$ consists of all pairs $((\alpha, y, m), (\beta, z, n))$ such that $\alpha \cdot \beta = 1$. One sees very quickly that the isomorphism $f_i \times f_i$ of $G(L^i) \times G(L^i)$ into $G(i\delta) \times G(i\delta)$ maps the kernel of μ_i onto the kernel of $\mu_{i\delta}$. Since μ_i maps $G(L^2) \times G(L^2)$ surjectively onto $G(M^i)$ and $\mu_{i\delta}$ maps $G(i\delta) \times G(i\delta)$ surjectively onto $G(i\delta, i\delta)$, the isomorphism $f_i \times f_i$ induces a theta structure g_i on $G(M^i)$ such that the diagram

$$(17.4) \qquad \begin{array}{ccc} G(L^i) \times G(L^i) & \xrightarrow{f_i \times f_i} & G(i\delta) \times G(i\delta) \\ \downarrow{\scriptstyle \mu_i} & & \downarrow{\scriptstyle \mu_{i\delta}} \\ G(M^i) & \xrightarrow{g_i} & G(i\delta, i\delta) \end{array}$$

is commutative. By equation 14.3 we have

$$3\bar{\}| = \mu_2 \circ \Delta_G.$$

It follows that the homomorphism

$$3\bar{\}| : G(L^2) \to G(M^2)$$

is transformed by f_2 and g_2 into the homomorphism

$$\mu_{2\delta} \circ \Delta_{2\delta} : G(2\delta) \to G(2\delta, 2\delta).$$

If $z = (\alpha, y, m)$ is an element of $G(2\delta)$, we have

(17.5) $$\mu_{2\delta}\circ\Delta_{2\delta}(z) = (\alpha^2, y, y, m, m) = A(z)$$

where A is defined by the first of the equations 8.2. Similarly, by equation 14.7, the homomorphism $3\bar{\}|' : G(L^2) \to G(M^2)$ is transformed by f_2 and g_2 into the homomorphism

$$\mu_{2\delta} \circ \Delta'_{2\delta} : G(2\delta) \to G(2\delta, 2\delta).$$

If z is an element of $G(2\delta)$, we have

(17.6) $$\mu_{2\delta}\circ\Delta'_{2\delta}(z) = (\alpha^2, y, -y, m, m^{-1}) = A'(z)$$

where A' is defined by the second of the equations 8.2. Similarly, using equations 14.9, we see that $h_{(s_1, id)}$ is transformed into S_1 as defined by the first of equations 8.1 and $h_{(s_2, id)}$ is transformed into S_2 as defined by the second of equations 8.1. Applying f_1, f_2, g_1 and g_2 to the commutative diagrams 13.2 and 13.3 and applying Proposition 16.19 and the results of this section, we see that the homomorphism $h_{(\xi,\beta)}$ is transformed into a homomorphism

$$T : G(2\delta, 2\delta)' \to G(\delta, \delta)$$

which makes Diagrams 17.7 and 17.8 below commute.

(17.7)
$$
\begin{array}{ccc}
G(2\delta) & \xrightarrow{\eta_{(x,\chi)}} & G(\delta) \\
\downarrow{\scriptstyle A} & & \downarrow{\scriptstyle S_1} \\
G(2\delta, 2\delta)' & \xrightarrow{T} & G(\delta, \delta)
\end{array}
$$

(17.8)
$$
\begin{array}{ccc}
G(2\delta) & \xrightarrow{\eta_{(x,\chi)}} & G(\delta) \\
\downarrow{\scriptstyle A'} & & \downarrow{\scriptstyle S_2} \\
G(2\delta, 2\delta)' & \xrightarrow{T} & G(\delta, \delta)
\end{array}
$$

By the results of §8, Diagrams 8.3 and 8.4 determine the homomorphism $T_{(x,\chi)}$ completely. Therefore, we have proved the following result.

Proposition (17.9):   *Let (X, L) be a strongly symmetric separably polarized abelian variety and let*

(17.10)
$$f_1 : \quad G(L) \to G(\delta)$$

$$f_2 : \quad G(L^2) \to G(2\delta)$$

be a pair of symmetric theta structures such that Diagram 16.21 is commutative. Let $M = L \boxtimes L$ and let

(17.11)
$$g_1 : \quad G(M) \to G(\delta, \delta)$$

$$g_2 : \quad G(M^2) \to G(2\delta, 2\delta)$$

be the theta structures which make Diagram 17.4 commutative. Then the following diagram is commutative.

(17.12)
$$
\begin{array}{ccc}
G(M^2) & \xrightarrow{g_2} & G(2\delta, 2\delta) \\
\downarrow{\scriptstyle \xi_X} & & \downarrow{\scriptstyle T_{(x,\chi)}} \\
G(M) & \xrightarrow{g_1} & G(\delta, \delta)
\end{array}
$$

Using f_1 and f_2 we can view $\Gamma(L)$ and $\Gamma(L^2)$ respectively as modules for $G(\delta)$ and $G(2\delta)$. By the results of §2, we can in fact identify $\Gamma(L)$ with $V(\delta)$ and $\Gamma(L^2)$ with $V(2\delta)$, the isomorphisms being unique up to nonzero scalar factors. Similarly, using g_1 and g_2, we can identify $\Gamma(M)$ with $V(\delta,\delta)$ and $\Gamma(M^2)$ with $V(2\delta, 2\delta)$, the isomorphisms being unique up to nonzero scalar factors. Applying the Mumford functor to the morphism (ξ, β) of $(X \times X, M^2)$ onto $(X \times X, M)$, where

$$\beta : M^2 \to \xi^* M$$

is the isomorphism of Lemma 12.6, we have

(17.13)
$$Mum((\xi, \beta)) = \left(G(M^2)'_{(\xi,\beta)}, h_{(\xi,\beta)}, \Omega_{(\xi,\beta)} \right)$$

as in §10. Here we recall from §10 that $h_{(\xi,\beta)}$ is a homomorphism of $G(M^2)'_{(\xi,\beta)}$ onto $G(M)$ and

$$\Omega_{(\xi,\beta)} : \Gamma(M) \to \Gamma(M^2)$$

is a k-linear mapping such that for all z in $G(M^2)'_{(\xi,\beta)}$ we have

(17.14)
$$\Omega_{(\xi,\beta)} \circ \rho_M \left(h_{(\xi,\beta)}(z) \right) = \rho_{M^2}(z) \circ \Omega_{(\xi,\beta)}.$$

Using g_1 and g_2 and the isomorphisms of $\Gamma(M^i)$ onto $V(i\delta, i\delta)$ given above for $i = 1, 2$, the morphism

(17.15)
$$\left(G(M^2)'_{(\xi,\beta)}, h_{(\xi,\beta)}, \Omega_{(\xi,\beta)} \right)$$

is transformed into the morphism

(17.16)
$$\left(G(2\delta, 2\delta)', T_{(x,\chi)}, \Omega' \right),$$

where Ω' is to be determined. By Corollary 7.11, the operator $\Omega_{(x,\chi)}$ is an isomorphism of the $G(\delta,\delta)$module $V(\delta,\delta)$ onto the subspace V of $V(2\delta,2\delta)$ consisting of all functions ψ on $K(2\delta,2\delta)$ which are fixed by every operator of the form $\rho_{(2\delta,2\delta)}(z)$ with z in the kernel of $T_{(x,\chi)}$. By Diagram 17.12 and the definition of the Mumford functor, the operator Ω' maps $V(\delta,\delta)$ isomorphically onto the space V and is also an isomorphism of $G(\delta,\delta)$-modules. Therefore Ω' and $\Omega_{(x,\chi)}$ differ only by a nonzero scalar factor. By modifying our isomorphism $\Gamma(M) \to V(\delta,\delta)$ by a suitable scalar factor if necessary, we may assume that $\Omega' = \Omega_{(x,\chi)}$. We have therefore proved the following result which we call the **Addition Formula**.

Theorem (17.17): *Retain the hypotheses of Proposition 17.9. For $i = 1, 2$ let*

$$\phi_i : \Gamma(M^i) \to V(i\delta, i\delta)$$

be an isomorphism of $\Gamma(M^i)$ onto $V(i\delta, i\delta)$ such that for all z in $G(M^i)$ we have

(17.18) $$\phi_i \circ \rho_{M^i}(z) = \rho_{i\delta}(z) \circ \phi_i.$$

Let

$$\Omega' : V(\delta,\delta) \to V(2\delta, 2\delta)$$

be the linear mapping given by

(17.19) $$\Omega' = \phi_2 \circ \Omega_{(\xi,\beta)} \circ \phi_1^{-1}$$

where $\Omega_{(\xi,\beta)}$ is the linear mapping of $\Gamma(M)$ into $\Gamma(M^2)$ given by

(17.20) $$\Omega_{(\xi,\beta)}(s) = \beta^{-1} \circ \xi^*(s)$$

and where

$$\beta : M^2 \to \xi^*(M)$$

is an isomorphism. Then up to a scalar factor, which after suitably normalizing ϕ_1 and ϕ_2 we may take to be 1, we have

(17.21) $$(\Omega'\Phi)(v,w) = \begin{cases} \chi(v) \cdot \Phi\left(j_\delta^{-1}(U), j_\delta^{-1}(V)\right) & \text{if } U \in 2K(2\delta) \\ 0 & \text{if not} \end{cases}$$

where $U = v + w + x$ and $V = v - w + x$.

The line bundle L^2 is also strongly symmetric. In fact, we have

(17.22) $$\begin{aligned} H(L^2) \cap X_2 &= X_2 \\ \epsilon_{L^2} &= (\epsilon_L)^2 \equiv 1. \end{aligned}$$

Therefore, applying Proposition 16.20 to the symmetric theta structure

$$f_2 : G(L^2) \to G(2\delta)$$

we see at once that there is a symmetric theta structure $f_4 : G(L^4) \to G(4\delta)$ such that the diagram

$$(17.23) \qquad \begin{array}{ccc} G(L^4) & \xrightarrow{f_4} & G(4\delta) \\ \downarrow{\eta'} & & \downarrow{\eta_{(0,1)}} \\ G(L^2) & \xrightarrow{f_2} & G(2\delta) \end{array}$$

is commutative, where the mapping η' is given by

$$(17.24). \qquad \eta' = \eta^{L^2}$$

We can use f_4 to view $\Gamma(L^4)$ as a $G(4\delta)$-module and choose an isomorphism ϕ_4 of the $G(4\delta)$-module $\Gamma(L^4)$ onto the $G(4\delta)$-module $V(4\delta)$. Denote by $[2]$ the linear mapping of $V(\delta)$ into $V(4\delta)$ corresponding to the mapping

$$2_X^* : \Gamma(L) \to \Gamma(L^4).$$

Since

$$(17.25) \qquad \xi_X \circ \xi_X = 2_X \times 2_X,$$

we have for Φ_1 and Φ_2 in $V(\delta)$ that

$$(17.26) \qquad [2]\Phi_1 \otimes [2]\Phi_2 = \Omega'' \circ \Omega' (\Phi_1 \otimes \Phi_2).$$

Here Ω'' denotes the linear mapping

$$\Omega'' : V(2\delta, 2\delta) \to V(4\delta, 4\delta)$$

of $V(2\delta, 2\delta)$ into $V(4\delta, 4\delta)$ obtained by replacing L by L^2, (x, χ) by $(0, 1)$, f_2 by f_4, f_1 by f_2, *et cetera* in the statement of Theorem 17.17. We then have

$$(17.27) \qquad (\Omega''\Phi)(u, v) = \begin{cases} \Phi\left(j_{2\delta}^{-1}(u+v), j_{2\delta}^{-1}(u-v) \right) & \text{if } u+v \in 2K(4\delta) \\ \\ 0 & \text{if not} \end{cases}$$

for all Φ in $K(2\delta)$. Therefore for all Φ_1 and Φ_2 in $V(\delta)$ and all u, v in $K(4\delta)$, we have

$$(17.28) \qquad \begin{aligned} ([2]\Phi_1 \otimes [2]\Phi_2)(u, v) &= \Omega'(\Phi_1 \otimes \Phi_2)\left(j_{2\delta}^{-1}(u+v), j_{2\delta}^{-1}(u-v) \right) \\ &= \chi(u+v) \cdot (\Phi_1 \otimes \Phi_2)\left(j_\delta^{-1}\left(x + j_{2\delta}^{-1}(2u) \right), j_\delta^{-1}\left(x + j_{2\delta}^{-1}(2v) \right) \right) \end{aligned}$$

if $u + v$ belongs to $2K(4\delta)$ and $x + j_{2\delta}^{-1}(2u)$ belongs to $2K(2\delta)$, and otherwise,

$$(17.29) \qquad ([2]\Phi_1 \otimes [2]\Phi_2)(u, v) = 0.$$

It follows that up to a scalar factor which can be normalized to be 1, we have

(17.30) $([2]\Phi)(u) = \begin{cases} \chi(u) \cdot \Phi(j_\delta^{-1}(x + j_{2\delta}^{-1}(2u))) & \text{if } x + j_{2\delta}^{-1}(2u) \in 2K(2\delta) \\ 0 & \text{if not} \end{cases}$

This is the **duplication formula**.

§18   The Fundamental Relation, Product Formula and Inversion Formula

We retain the notation and assumptions of the preceding section. Choose an isomorphism $\lambda_0^{(1)}$ of $L(0)$ onto k, where $L(0)$ denotes the tensor product of the fibre L_0 of L are 0 with the residue class field k of $\mathcal{O}_{X,0}$, the tensor product being taken with respect to $\mathcal{O}_{X,0}$. Then $\lambda_0^{(1)}$ induces

(18.1) $$\lambda_0^{(2)} : L^2(0) \to k$$

and

(18.2) $$\lambda_0^{(4)} : L^4(0) \to k.$$

Using the isomorphisms

$$\phi_i : \Gamma(L^i) \to V(i\delta)$$

of $\Gamma(L^i)$ onto $V(i\delta)$ for $i = 1, 2, 4$, we define the function

$$q_{L^i} : K(i\delta) \to k$$

by the rule

(18.3) $$\lambda_0^{(i)}[t(0)] = \sum_{z \in K(i\delta)} (\phi_i t)(z) \, q_{L^i}(z).$$

We may also describe the function q_{L^i} in the following way: let $x \in K(i\delta)$. Let $\nabla_x : K(i\delta) \to k^\times$ be the function which is 1 at x and 0 elsewhere. Then $q_{L^i}(x)$ is the value at 0 of the section t_x of $\Gamma(L^i)$, where $\phi_i(t_x) = \nabla_x$, i.e. $q_{L^i}(x) = \lambda_0^{(i)}(t_x)$.

The isomorphism $\lambda_0^{(1)}$ also induces

$$\mu_0^{(i)} : M^i(0) \to k$$

for $i = 1, 2, 4$ and we have

(18.4) $$\mu_0^{(i)}[t(0)] = \sum_{x,y \in K(i\delta)} ((\phi_i \otimes \phi_i)(t)) \, q_{L^i}(x) \, q_{L^i}(y)$$

for all t in $\Gamma(X \times X, M^i)$. There is a constant κ such that

(18.5) $$\mu_0^{(2)}[(\Omega_{(\xi,\beta)}t)(0)] = \kappa \mu_0^{(1)}(t(0))$$

for all t in $\Gamma(X \times X, M^i)$. Replacing β by $\kappa^{-1}\beta$, we may assume that $\kappa = 1$. Therefore for all f in $V(\delta,\delta)$, we have

(18.6) $$\sum_{a,b \in K(\delta)} f(a,b) q_L(a) \cdot q_L(b) = \sum_{u,v \in K(2\delta)} \left(\Omega_{(\xi,\beta)} f \right)(u,v) \cdot q_{L^2}(u) \cdot q_{L^2}(v).$$

In particular, take $f(a,b)$ to be the delta function $\delta_{(w,z)}$ at (w,z). Then we have

(18.7) $$q_L(w) \cdot q_L(z) = \sum \chi(u) \, q_{L^2}(u) \cdot q_{L^2}(v)$$

where the summation runs over all u and v in $K(2\delta)$ such that

(18.8) $$j_\delta(w) = u + v + x$$

and

(18.9) $$j_\delta(z) = u - v + x.$$

We will refer to equation 18.7 as the **fundamental relation among the theta constants**. In particular cases, some of which we will examine in §§19-22, it is possible to eliminate the terms involving q_{L^2} in the fundamental relations in order to obtain relations among the q_L's alone. However, there does not appear to be a simple uniform way of doing so and the simplest general relation which can be given is 18.7. We can, however, make the following useful observation.

Lemma (18.10): *The function q_L is even or odd according to whether the Arf invariant of L is 1 or -1.*

Proof: Choose z in $K(\delta)$ such that $q_L(z)$ is nonzero. Then for w in $K(\delta)$, the value of $q_L(z) \cdot q_L(-w)$ is obtained by interchanging u and v in the right hand side of 18.7. Since

$$u - v + x = j_\delta(z)$$

lies in $2K(2\delta)$, it follows that

$$\chi(v) = \chi(u + x).$$

Substituting this value for $\chi(v)$ and using the fact that $q_L(z) \neq 0$, we see that

$$q_L(-w) = \chi(x) \cdot q_L(w).$$

We can now describe explicitly the linear mapping of $\Gamma(L) \otimes \Gamma(L)$ into $\Gamma(L^2)$ determined by the tensor product of sections of L. This is accomplished by means of the identity

(18.11) $$\xi_X \circ s_1 = \Delta.$$

If t and t' are sections of L then $t \boxtimes t'$ is a section of M and $\Delta^*(t \boxtimes t')$ is simply

$t \otimes t'$ which is what we want to compute.

If g_1 and g_2 belong to $\Gamma(L^2)$ then letting $g = g_1 \otimes g_2$, we have

(18.12)
$$s_1^*(g) = (\lambda_0^{(2)}[g_2(0)]) \cdot g_1.$$

In case $\phi_2(g_2) = f$ in $V(2\delta)$, we have

(18.13)
$$\lambda_0^{(2)}[g_2(0)] = \sum_{y \in K(2\delta)} f(y) \, q_{L^2}(y).$$

Therefore

(18.14)
$$\begin{aligned}
s_1^*(g) &= \left(\lambda_0^{(2)}[g_2(0)] \right) \cdot g_1 \\
&= \left(\sum_{y \in K(2\delta)} f(y) \, q_{L^2} \right) g_1.
\end{aligned}$$

Denote by
$$\sigma_1 : V(2\delta, 2\delta) \to V(2\delta)$$
the linear mapping from $V(2\delta, 2\delta)$ to $V(2\delta)$ corresponding the the linear mapping
$$s_1^* : \Gamma(M^2) \to \Gamma(L^2).$$
Then since sections of $\Gamma(M^2)$ of the form $g_1 \otimes g_2$ span $\Gamma(M^2)$, we conclude that for all g in $V(\delta, \delta)$ we have

(18.15)
$$(\sigma_1 g)(w) = \sum_{y \in K(2\delta)} g(w, y) \cdot q_{L^2}(y).$$

On the other hand, we know from Theorem 17.17 that $\Omega_{(\xi, \beta)}$ corresponds to $\Omega_{(x, \chi)}$. Therefore if we denote by
$$* : V(2\delta, 2\delta) \to V(2\delta)$$
the operator corresponding to the multiplication
$$\Gamma(L) \otimes \Gamma(L) \to \Gamma(L^2)$$
we have for all f_1 and f_2 in $V(\delta)$ that

(18.16)
$$(f_1 * f_2)(u) = \sum_{y \in K(2\delta)} \Omega_{(x, \chi)}(f_1 \otimes f_2)(u, y) \cdot q_{L^2}(y)$$

We may rewrite this as

(18.17) $(f_1 * f_2)(u) = \chi(u) \displaystyle\sum_{z \in K(2\delta)}{}' f_1\left(j_\delta^{-1}(u + y + x) \right) \cdot f_2\left(j_\delta^{-1}(u - y + x) \right) \cdot q_{L^2}(y)$

where the summation runs over all y in $K(2\delta)$ such that $u + y + x$ lies in $2K(2\delta)$.

We can also write equations 18.16 and 18.17 in the form

$$(18.18) \quad (f_1 * f_2)(u) = \chi(u) \sum_{z \in K(\delta)} f_1(z) \cdot f_2 \left(j_\delta^{-1}(2u) - z \right) \cdot q_{L^2} \left(j_\delta(z) - u + x \right)$$

by writing $y = x - u + j_\delta(z)$ and letting z run over $K(\delta)$. Then equation 18.16 or, what is the same, equation 18.18 is the **product formula**.

If instead of equation 18.11 we use the formula

$$\xi_\chi \circ s_2 = \Delta',$$

we obtain the formula

$$(18.19) \qquad\qquad s_2^*(g) = \lambda_0^{(2)}[g_1(0)] \cdot g_2$$

and if

$$\sigma_2 : V(2\delta, 2\delta) \to V(2\delta)$$

is the linear mapping corresponding to

$$s_2^* : \Gamma(M^2) \to \Gamma(L^2),$$

we have

$$(18.20) \qquad\qquad (\sigma_2 g)(w) = \sum_{y \in K(2\delta)} g(y, w) \cdot q_{L^2}(y).$$

On the other hand, using the normalized isomorphism of L onto $\iota^*(L)$, we obtain an involution, which we will denote by ι^*, on the space $\Gamma(L)$. Using the isomorphism ϕ_1 of $\Gamma(L)$ onto $V(\delta)$, the involution ι^* corresponds to an involution of $V(\delta)$ which we will denote by $[-1]$. If w is an element of X, we have

$$(18.21) \qquad p_2 \circ \xi \circ s_2(w) = p_2 \circ \xi(0, w) = p_2(w, -w) = -w = \iota(w),$$

so

$$p_2 \circ \xi \circ s_2 = \iota.$$

Similarly, we have

$$p_2 \circ \xi \circ s_2(w) = 1_X.$$

Therefore, if s, t belong to $\Gamma(L)$, we have

$$(18.22) \qquad\qquad s_2^* \xi^* \left(s \boxed{\times} t \right) = s \otimes \iota^* t.$$

It follows that if for f, f' in $V(\delta)$ we put

$$g(u, y) = f(u) \cdot f'(y),$$

then

$$(f * [-1]f')(u) = \sum_{y \in K(2\delta)} (\Omega_{(x,\chi)}g)(y, u) \cdot q_{L^2}(y)$$

(18.23)

$$= \sum_{y \in K(2\delta)}' \chi(y) \cdot f\left(j_\delta^{-1}(y + u + x)\right) \cdot f'\left(j_\delta^{-1}(y - u + x)\right) \cdot q_{L^2}$$

where the summation runs over all y in $K(\delta)$ such that $y + u + x$ lies in $2K(2\delta)$. We can rewrite equation 18.23 as

(18.24)

$$(f * [-1]f')(u) = \chi(u) \cdot \sum' f\left(j_\delta^{-1}(y + u + x)\right) \cdot f'\left(j_\delta^{-1}(y - u + x)\right) \cdot q_{L^2}(y)$$

$$= f * f''(u)$$

where $f''(u) = \chi(x) \cdot f'(-u)$ for all u in $K(\delta)$. Since the homogeneous coordinate ring of (X, L) is an integral domain, we conclude that

(18.25)
$$[-1]f(w) = \chi(x)f(-w)$$

for all f in $V(\delta)$. This is the **inversion formula**.

Chapter IV

Geometry and Arithmetic of the Fundamental Relations

§19 Special Cases of the Fundamental Relation Among Theta Constants

In this section we will examine the fundamental relation 18.7 in certain special cases. Retaining the notation and assumptions of that section, let us suppose that the order of the group $H(L)$ is an odd integer. If z, w belong to $K(\delta)$ and u, v are elements of $H(2\delta)$ such that

(19.1)
$$u + v + x = j_\delta(z)$$
$$u - v + x = j_\delta(w)$$

then we have

$$2u = j_\delta(z + w)$$

and

$$2v = j_\delta(z - w).$$

Since $H(L)$ has odd order, we can find u_0 and v_0 in $K(\delta)$ such that

$$2u_0 = z + w$$

and

$$2v_0 = z - w$$

Therefore the fundamental relation can be rewritten as
(19.2)
$$q_L(z)\, q_L(w) = \sum_{\substack{a, b \in K(2\delta)_2 \\ a + b = x}} \chi\left(j_\delta(u_0) + a\right) \cdot q_{L^2}\left(j_\delta(u_0) + a\right) \cdot q_{L^2}\left(j_\delta(v_0) + b\right)$$

$$= \sum_{a \in K(2\delta)_2} \chi(a) \cdot q_{L^2}\left(j_\delta(u_0) + a\right) \cdot q_{L^2}\left(j_\delta(v_0) + a + x\right)$$

whenever $H(L)$ has odd order. Here we have used the fact that $j_\delta(u_0)$ lies in $2K(2\delta)$

so that

$$\chi(j_\delta(u_0)) = 1.$$

Let

$$B = \{a \in K(2\delta)_2 \mid \chi(a) = 1\}$$

be the set of all elements a of $K(2\delta)_2$ such that $\chi(a) = 1$. We then have two cases, depending on the Arf invariant of L.

Case 1: L Has Nontrivial Arf Invariant

Since the Arf invariant of L is $\chi(x)$, it is the same to say that we are dealing with the case in which

$$\chi(x) = -1.$$

Then Equation 19.2 becomes

$$
\begin{aligned}
(19.3) \qquad q_L(z) \cdot q_L(w) \;=\; & \sum_{a \in B} q_{L^2}\left(j_\delta(u_0) + a\right) \cdot q_{L^2}\left(j_\delta(v_0) + a + x\right) \\
& - \sum_{a \in B} q_{L^2}\left(j_\delta(u_0) + a + x\right) \cdot q_{L^2}\left(j_\delta(v_0) + a\right)
\end{aligned}
$$

and by Lemma 18.10, q_L is an odd function on $K(\delta)$. If we combine the two summations we therefore obtain

$$(19.4) \qquad q_L(z) \cdot q_L(w) = \sum_{a \in B} \begin{vmatrix} q_{L^2}(j_\delta(u_0) + a) & q_{L^2}(j_\delta(v_0) + a) \\ q_{L^2}(j_\delta(u_0) + a + x) & q_{L^2}(j_\delta(v_0) + a + x) \end{vmatrix}$$

where

$$z = u_0 + v_0$$

and

$$w = u_0 - v_0$$

and where u_0 and v_0 lie in $K(\delta)$. It is the same to write

$$(19.5) \qquad q_L(u + v) \cdot q_L(u - v) = \sum_{a \in B} \begin{vmatrix} q_{L^2}(j_\delta(u) + a) & q_{L^2}(j_\delta(v) + a) \\ q_{L^2}(j_\delta(u) + a + x) & q_{L^2}(j_\delta(v) + a + x) \end{vmatrix}$$

for all u and v in $K(\delta)$. For example, suppose that the dimension of X is equal to 1. Then we have $B = (0)$ and

$$(19.6) \qquad q_L(u + v) \cdot q_L(u - v) = \begin{vmatrix} q_{L^2}(j_\delta(u)) & q_{L^2}(j_\delta(v)) \\ q_{L^2}(j_\delta(u) + x) & q_{L^2}(j_\delta(v) + x) \end{vmatrix}$$

Let u, v, w, z be any elements of $K(\delta)$. Applying the Plücker relations to the right hand side of 19.6 we obtain the following quartic relation.

$$0 \;=\; q_L(u+v) \cdot q_L(u-v) \cdot q_L(w+z) \cdot q_L(w-z)$$

(19.7)
$$+\; q_L(u+w) \cdot q_L(u-w) \cdot q_L(z+v) \cdot q_L(z-v)$$

$$+\; q_L(u+z) \cdot q_L(u-z) \cdot q_L(v+w) \cdot q_L(v-w)$$

Write V^+ and V^- to denote respectively the spaces

$$V^+ = \{f : K(\delta) \to k \mid (\forall u)\; f(u) = f(-u)\}$$

$$V^- = \{f : K(\delta) \to k \mid (\forall u)\; (f(u) = -f(-u))\}$$

of even and odd functions on $K(\delta)$. Let Λ denote the subspace

$$\Lambda = \{f : V(\delta,\delta) \to k \mid (\forall u)(\forall v)\; (f(u,v) = -f(v,u) \text{ and } f(-u,v) = f(u,v))\}$$

of $V(\delta,\delta)$ consisting of all functions f on $K(\delta,\delta) = K(\delta) \times K(\delta)$ such that for all u, v in $K(\delta)$ we have $f(u,v) = -f(v,u)$ and $f(-u,v) = f(u,v)$. Let Σ denote the subspace

$$\Sigma = \{g : V(\delta,\delta) \to k \mid (\forall u)(\forall v)\; (g(u,v) = g(v,u) \text{ and } g(-u,v) = -g(u,v))\}$$

of $V(\delta,\delta)$ consisting of all functions g on $K(\delta,\delta)$ such that for all u, v in $K(\delta)$ we have $g(u,v) = g(v,u)$ and $g(-u,v) = -g(u,v)$. If h is any function on $K(\delta,\delta)$, denote by $T(h)$ the function on $K(\delta,\delta)$ defined by

(19.8)
$$T(h)\,(u,v) = h\left(\frac{u+v}{2}, \frac{u-v}{2}\right)$$

Then T is an automorphism of the vector space $V(\delta,\delta)$. Furthermore one can easily show that T induces an isomorphism of Λ onto Σ. If α, β belong to $V(\delta)$ denote by $\alpha \wedge \beta$ and $\alpha \vee \beta$ the functions on $K(\delta,\delta)$ defined respectively by

(19.9)
$$(\alpha \wedge \beta)\,(u,v) \;=\; \alpha(u)\beta(v) - \alpha(v)\beta(u) \;=\; \begin{vmatrix} \alpha(u) & \alpha(v) \\ \beta(u) & \beta(v) \end{vmatrix}$$

$$(\alpha \vee \beta)\,(u,v) \;=\; \alpha(u)\beta(v) + \alpha(v)\beta(u).$$

It is straightforward to verify that if α and β are both even functions on $K(\delta)$ then $\alpha \wedge \beta$ will belong to the space Λ, while if α and β are both odd then $\alpha \vee \beta$ will belong to the space Σ. The expressions $\alpha \wedge \beta$ and $\alpha \vee \beta$ are bilinear in α and β. Furthermore, one sees very easily that $\alpha \wedge \beta$ is alternating in α and β and $\alpha \vee \beta$ is symmetric. Denote by $\wedge$ and $\vee$ the linear mappings

(19.10)
$$\wedge \ : \ \textstyle\bigwedge^2(V^+) \to \Lambda$$
$$\vee \ : \ Sym^2(V^-) \to \Sigma$$

determined by the pairings $\wedge$ and $\vee$ respectively. They are easily seen to be isomorphisms. It follows that there is one and only one isomorphism

$$\tau : \overset{2}{\textstyle\bigwedge}(V^+) \to Sym^2(V^-)$$

of $\bigwedge^2(V^+)$ onto $Sym^2(V^-)$ such that

$$\vee \circ \tau = \mathcal{T} \circ \wedge.$$

The group $GL(V^+)$ acts on $\bigwedge^2(V^+)$ and therefore on Λ via $\vee$. There is one and only one closed orbit for $GL(V^+)$ in $\mathbf{P}(\Lambda)$, namely the image under $\wedge$ of the Grassmanian of all 2-planes in V^+. We will denote that closed orbit by Gr. By means of the isomorphism $\mathcal{T}$ of Λ onto Σ, we may identify Gr with a subvariety of $\mathbf{P}(\Sigma)$.

The group $GL(V^-)$ acts on $Sym^2(V^-)$ and therefore on Σ via $\wedge$. There is one and only one closed orbit for $GL(V^-)$ in $\mathbf{P}(\Sigma)$, namely the image of $\mathbf{P}(V^-)$ in $\mathcal{P}(\Sigma)$ under the mapping $[h] \mapsto [h \vee h]$. Denote[4] the image of this mapping by $\mathcal{V}$. Since $\mathcal{V}$ is a subvariety of $\mathbf{P}(\Sigma)$ and since Gr has been identified with a subvariety of $\mathbf{P}(\Sigma)$, we can consider the intersection of $\mathcal{V}$ and Gr. Denote by $\mathcal{L}$ the preimage in $\mathbf{P}(V^-)$ of $\mathcal{V} \cap Gr$ under the isomorphism of $\mathbf{P}(V^-)$ onto $\mathcal{V}$ induced by $\vee$. Then $\mathcal{L}$ is the locus of all $[h]$ in $\mathbf{P}(V^-)$ such that for some α, β in V^+ we have

(19.11)
$$h \vee h = \mathcal{T}(\alpha \wedge \beta).$$

It is the same to say that $[h]$ lies in $\mathcal{L}$ if and only if for some α, β in V^+ we have

(19.12)
$$h(u + v) \cdot h(u - v) = \begin{vmatrix} \alpha(u) & \alpha(v) \\ \beta(u) & \beta(v) \end{vmatrix}$$

for all u, v in $K(\delta)$. In particular, if q_L is not identically 0 the point $[q_L]$ of $\mathbf{P}(V^-)$ lies in $\mathcal{L}$. Applying the Plücker relations to 19.12 we obtain

(19.13)
$$
\begin{aligned}
0 \ = \ & h(u + v) \cdot h(u - v) \cdot h(w + z) \cdot h(w - z) \\
+ \ & h(u + w) \cdot h(u - w) \cdot h(z + v) \cdot h(z - v) \\
+ \ & h(u + z) \cdot h(u - z) \cdot h(v + w) \cdot h(v - w)
\end{aligned}
$$

---

[4] This notation, in which $\mathcal{V}$ is the initial of the name Veronese, is motivated by the case of the Veronese surface in $\mathbf{P}^5$, which is a special case of this construction. In general, the variety $\mathcal{V}$ would be referred to as a 2-uply embedded projective space and the embedding $[h] \mapsto [h \vee h]$ as the 2-uple embedding.

for every $[h]$ in $\mathcal{L}$ and all u, v, w, z in $K(\delta)$. Conversely, suppose that $[h]$ is a point of $\mathbf{P}(V^-)$ such that 19.13 holds for all u, v, w, z in $K(\delta)$. Then h does not vanish identically so we can choose an element t in $K(\delta)$ such that $h(t) \neq 0$. Let $\alpha(u) = c \cdot h(u)^2$ where c is a constant to be chosen later and let $\beta(u) = h(t+u) \cdot h(t-u)$. Then the function $\alpha \wedge \beta$ is given by

$$(19.14) \qquad (\alpha \wedge \beta) = c \cdot \left(h(r)^2 \cdot h(t+s) \cdot h(s-t) - h(s)^2 \cdot h(t+r) \cdot h(t-r) \right).$$

If we take $u = 0$, $v = r$, $w = s$ and $z = t$ in 19.13 we obtain

$$0 = h(r) \cdot h(-r) \cdot h(s+t) \cdot h(s-t)$$

$$(19.15) \qquad\qquad + h(s) \cdot h(-s) \cdot h(t+r) \cdot h(t-r)$$

$$+ h(t) \cdot h(-t) \cdot h(r+s) \cdot h(r-s).$$

Using the fact that h is an odd function, we conclude that

$$(19.16) \qquad\qquad (\alpha \wedge \beta)(r,s) = c \cdot h(t)^2 \cdot h(r+s) \cdot h(r-s).$$

If we now take $c = h(t)^{-2}$, we see that

$$h(r+s)\, h(r-s) = (\alpha \wedge \beta)(r,s) = \begin{vmatrix} \alpha(r) & \alpha(s) \\ \beta(r) & \beta(s) \end{vmatrix}$$

for all r, s in $K(\delta)$, which proves that h lies in $\mathcal{L}$. We have therefore proved the following result.

Theorem (19.17):   *The locus defined by 19.13 is the same as $\mathcal{L}$. In particular, the fundamental relation 18.7 in the form of 19.6 for $n = 1$ says that the point $[q_L]$ of $\mathbf{P}(V^-)$ lies in the intersection of a Veronese variety and a Grassmannian, or what is the same, in the intersection of the minimal orbits for $GL(V^-)$ and $GL(V^+)$ in $\mathbf{P}(\Sigma)$.*

For $n > 1$, the fundamental relation becomes 19.5 and implies that $[q_L]$ lies in the intersection of $\mathcal{V}$ with a suitable secant locus of the Grassmannian. It would be very interesting to study the geometry more closely.

Case 2: L has trivial Arf invariant

We are dealing with the case where $\chi(x) = 1$. Then Equation 19.2 becomes

$$q_L(z) \cdot q_L(w) = \sum_{a \in B} q_{L^2}\left(j_\delta(u_0) + a \right) \cdot q_{L^2}\left(j_\delta(v_0) + a + x \right)$$

$$(19.18)$$

$$- \sum_{a \in B} q_{L^2}\left(j_\delta(u_0) + a + y \right) \cdot q_{L^2}\left(j_\delta(v_0) + a + x + y \right)$$

where y is an element of $K(2\delta)_2$ such that

$$\chi(y) = -1.$$

Here we are assuming that χ is a nontrivial character. If instead χ is trivial, then 19.18 should be replaced by the simpler expression

$$(19.19) \qquad q_L\left(z\right)\cdot q_L\left(w\right) = \sum_{a\in K(2\delta)_2} q_{L^2}\left(j_\delta(u_0)+a\right)\cdot q_{L^2}\left(j_\delta(v_0)+a+x\right).$$

Whether χ is trivial or not, the function q_L is, according to Lemma 18.10, an even function. In practice we will be interested in the case in which $\chi = 1$ and $x = 0$. When that is the case, we have

$$(19.20) \qquad q_L\left(z\right)\cdot q_L\left(w\right) = \sum_{a\in K(2\delta)_2} q_{L^2}\left(j_\delta(u_0)+a\right)\cdot q_{L^2}\left(j_\delta(v_0)+a\right)$$

where

$$z = u_0 + v_0$$

and

$$w = u_0 - v_0.$$

It is the same to write

$$(19.21) \qquad q_L\left(u+v\right)\cdot q_L\left(u-v\right) = \sum_{a\in K(2\delta)_2} q_{L^2}\left(j_\delta(u)+a\right)\cdot q_{L^2}\left(j_\delta(v)+a\right)$$

for all u,v in $K(\delta)$. Let us write d to denote the order of $K(\delta)$ and let g denote the dimension of X. Let

$$S : K(\delta) \times K(\delta) \to k$$

denote the $d \times d$ matrix indexed by the elements of $K(\delta) \times K(\delta)$ in which the entry $S(u,v)$ corresponding to the element (u,v) of $K(\delta) \times K(\delta)$ is given by

$$(19.22) \qquad S(u,v) = q_L\left(u+v\right)\cdot q_L\left(u-v\right).$$

On the other hand, let

$$T : K(2\delta)_2 \times K(\delta) \to k$$

denote the $2^g \times d$ matrix indexed by the elements of $K(2\delta)_2 \times K(\delta)$ in which the entry $t(a,u)$ corresponding to the element (a,u) of $K(2\delta)_2 \times K(\delta)$ is given by

$$(19.23) \qquad t(a,u) = q_{L^2}\left(j_\delta(u)+a\right).$$

Then equation 19.22 can be written in the simple form

$$(19.24) \qquad S = {}^tT \cdot T.$$

The following result is now obvious:

Proposition (19.25):   *S is a symmetric matrix of rank $\leq 2^g$.*

Remark (19.26):   It is known in some cases (cf. [Ko1],[Ko2],[Sek]) that the matrix T has rank equal to 2^g. However, we do not know the rank of S in general.

§20   The Geometry of the Modular Curve Via the Fundamental Relation

In this section, p will denote a prime number which is greater than or equal to 7. We will be concerned with the locus $\mathcal{L}$ which we introduced in the discussion of Case 1 in §19. We will retain the notation of that discussion and we will take

$$\delta = (p),$$

so that

$$K(\delta) = \mathbf{Z}/p\mathbf{Z}.$$

We will assume that the characteristic of k does not divide the order of $PSL_2(\mathbf{F}_p)$. For every element t of $K(\delta)$, let E_t denote the linear form on V^- given by $E_t(h) = h(t)$. If w, x, y, z are any elements of $K(\delta)$, denote by $\Phi_{w,x,y,z}$ the quartic form given by

$$
\begin{aligned}
\Phi_{w,x,y,z} \;=\; & E_{w+x} \cdot E_{w-x} \cdot E_{y+z} \cdot E_{y-z} \\[4pt]
(20.1) \qquad\qquad + \;& E_{w+y} \cdot E_{w-y} \cdot E_{z+x} \cdot E_{z-x} \cdot \\[4pt]
+ \;& E_{w+z} \cdot E_{w-z} \cdot E_{x+y} \cdot E_{x-y}
\end{aligned}
$$

Then $\mathcal{L}$ is defined by the equations

$$\Phi_{w,x,y,z} = 0$$

with w, x, y, z in $\mathbf{Z}/p\mathbf{Z}$. Since $E_{-t} = -E_t$ for every t in $K(\delta)$, we have $\Phi_{-w,x,y,z} = \Phi_{w,x,y,z}$. Furthermore, we have

$$(20.2) \qquad\qquad \Phi_{x,w,y,z} = \Phi_{x,y,z,w} = -\Phi_{w,x,y,z}.$$

Since the odd permutations

$$(20.3) \qquad\qquad \begin{pmatrix} 1234 \\ 2134 \end{pmatrix} \quad \text{and} \quad \begin{pmatrix} 1234 \\ 2341 \end{pmatrix}$$

generate the group of all permutations on four objects, it follows that if $(a, b, c, d) = \sigma(w, x, y, z)$ is a permutation of (w, x, y, z) then

$$(20.4) \qquad\qquad \Phi_{a,b,c,d} = (-1)^{\sigma} \Phi_{w,x,y,z},$$

where $(-1)^{\sigma}$ denotes the sign of the permutation σ. It follows from this and from the sentence preceding (20.2) that for all choices of signs we have

$$(20.5) \qquad\qquad \Phi_{\pm w, \pm x, \pm y, \pm z} = \Phi_{w,x,y,z}.$$

Furthermore, if two of the elements w, x, y, z are equal then $\Phi_{w,x,y,z} = 0$. It follows that each quartic $\Phi_{w,x,y,z}$ is equal, up to a sign, to a quartic $\Phi_{a,b,c,d}$ with

$$(20.6) \qquad\qquad 0 \le a < b < c < d \le \frac{p-1}{2}.$$

The locus $\mathcal{L}$ is therefore defined by $\binom{m}{4}$ quartics, where

$$2m - 1 = p.$$

These quartics are obtained from the $\binom{m}{4}$ quadrics containing the Grassmannian Gr. In general, these quartics are not distinct. For example, when $p = 11$, the cardinality of $\mathcal{Y}$ is 15 but there are only 10 distinct quartics, namely Φ_{0123}, $\Phi_{0124} = \Phi_{2345}$, $\Phi_{0125} = \Phi_{1234}$, Φ_{0134}, $\Phi_{0135} = \Phi_{1245}$, Φ_{0145}, $\Phi_{0234} = \Phi_{1345}$, Φ_{0235}, Φ_{0245}, $\Phi_{0345} = \Phi_{1235}$. When $p = 13$, the cardinality of $\mathcal{Y}$ is 35 but there are only 21 distinct quartics, namely Φ_{0123}, $\Phi_{0124} = \Phi_{3456}$, $\Phi_{0125} = \Phi_{1234}$, $\Phi_{0126} = \Phi_{2345}$, Φ_{0134}, $\Phi_{0135} = \Phi_{2356}$, $\Phi_{0136} = \Phi_{1245}$, Φ_{0145}, $\Phi_{0146} = \Phi_{1256}$, Φ_{0156}, $\Phi_{0234} = \Phi_{2456}$, Φ_{0235}, $\Phi_{0236} = \Phi_{1346}$, $\Phi_{0245} = \Phi_{1356}$, Φ_{0246}, Φ_{0256}, $\Phi_{0345} = \Phi_{1236}$, Φ_{0346}, $\Phi_{0356} = \Phi_{1246}$, $\Phi_{0456} = \Phi_{1345}$, $\Phi_{1235} = \Phi_{1456} = \Phi_{2346}$. Nevertheless, they are not all necessary to define the locus $\mathcal{L}$. Indeed, the following argument, due to Ramanan, shows that one can manage with a smaller space of quartics than that spanned by the Φ_{abcd} with (a, b, c, d) in $\mathcal{Y}$.

Theorem (Ramanan) (20.7): The locus $\mathcal{L}$ can be defined by $\binom{(p-1)/2}{3}$ quartics.

Proof: Let us start with V^-, the space of odd functions on $K(\delta)$. If $\phi \in V^-$, then the function ϕ^2, namely $x \mapsto (\phi(x))^2$, belongs to V^+. Let us now assume that ϕ is in the locus $\mathcal{L}$. In other words, we assume that there exist $F, G \in V^+$ such that

$$(20.8) \qquad \phi(x + y)\phi(x - y) = (F \wedge G)(x, y) = F(x)G(y) - G(x)F(y).$$

Now take $y = 0$ and conclude that ϕ^2 is a linear combination of F and G. Hence one might as well replace F by ϕ^2. Thus $\phi(x + y)\phi(x - y)$ is a decomposable tensor if and only if it is of the form $(\phi^2) \wedge G$ for some $G \in V^+$. But then for any ϕ whatever, denoting by f_0 the Dirac function at 0, we have that the function

$$(20.9) \qquad R(x, y) = \phi(x + y)\phi(x - y) + (f_0 \wedge \phi^2)(x, y)$$

vanishes whenever $x = 0$ or $y = 0$. This means that if we decompose $\Lambda^2(V^+)$ as a direct sum of $kf_0 \otimes V_0^+$ and $\Lambda^2(V_0^+)$, where V_0^+ is the space of even functions vanishing at 0, then R belongs to $\lambda^2(V_0^+)$. Hence, to say that $\phi(x + y)\phi(x - y)$ is a decomposable tensor is equivalent to saying that R is of the form $\phi^2 \wedge H$, where H belongs to V_0^+. The necessary and sufficient condition for this to happen is $R \wedge \phi^2 = 0$, which leads to $\binom{(p-1)/2}{3}$ equations.

For p=13, this number is 20, which is smaller than the number of distinct (hence linearly independent) quartics Φ_{abcd} with (a, b, c, d) in $\mathcal{Y}$. As Ramanan has remarked, one can actually reduce the number further, namely to $(p - 1)/2$, but as this does not lead to explicit equations, we will not discuss it here.

For every $t \neq 0$ in $K(\delta)$, let h_t denote the element of V^- defined by

$$(20.10) \qquad h_t(x) = \begin{cases} 1 & \text{if } x = t \\ -1 & \text{if } x = -t \\ 0 & \text{if } x \neq \pm t, \end{cases}$$

and let $\kappa_t = [h_t]$ be the point of $\mathbf{P}(V^-)$ determined by h_t. Then we clearly have

$$h_{-t} = -h_t$$

and

$$\kappa_{-t} = \kappa_t$$

for all $t \neq 0$ in $K(\delta)$.

Proposition (20.11):   *Let $\kappa = [h]$ be a point of $\mathcal{L}$. If*

$$h(s) = 0$$

for some $s \neq 0$ in $K(\delta)$ then

$$\kappa = \kappa_t$$

for some $t \neq 0$ in $K(\delta)$. Here the characteristic of k may equal 2.

Proof: Let

$$Z = \{x \in K(\delta) \mid h(x) = 0\}$$

and let

$$Z' = \{x \in K(\delta) \mid h(x) \neq 0\}$$

denote the complement of Z in $K(\delta)$. Since h is odd, $Z = -Z$ and by hypothesis $Z \neq \{0\}$. If we put $y = 0$ and $z = w + x$ we find that 20.1 becomes

$$(20.12) \quad 0 = -h(w+x)^3 \cdot h(w-x) + h(w)^3 \cdot h(w+2x) - h(x)^3 \cdot h(2w+x).$$

If $h(x) = 0$ then for all w in $K(\delta)$ we have

$$(20.13) \qquad h(w)^3 \cdot h(w+2x) = h(w+x)^3 \cdot h(w-x).$$

In particular,

$$h(w) \cdot h(w+2x) = 0$$

if and only if

$$h(w+x) \cdot h(w-x) = 0.$$

Since $h(0) = 0$ we certainly have

$$h(0) \cdot h(2x) = 0.$$

It follows by induction that

$$h(rx) \cdot h((r+2)x) = 0$$

for every positive integer r. Since p is prime, x generates $K(\delta)$. Consequently every element w of $K(\delta)$ is of the form rx for some positive integer r and we have

$$(20.14) \qquad h(w) \cdot h(w+2x) = 0$$

for all w in $K(\delta)$. Therefore, if w lies in Z' then $w + 2x$ does not. Since x was an arbitrary nonzero element of Z we conclude that if a, b lie in Z' and $a \neq b$ then

$(a - b)/2$ lies in Z'. Furthermore if a, b are elements of Z' such that $a \neq -b$ then since $-b$ lies in Z' we see that $(a + b)/2$ lies in Z'. If κ is not of the form κ_t then we can find a, b in Z' such that $a \neq b$ and $a \neq -b$. Then $(a + b)/2$ and $(a - b)/2$ lie in Z' and $(a + b)/2 \neq (a - b)/2$. Therefore

$$(20.15) \qquad \frac{\frac{a+b}{2} + \frac{a-b}{2}}{2} = \frac{a}{2}$$

lies in Z'. It follows that if κ is not of the form κ_t then

$$\frac{1}{2}Z' = Z',$$

or what is the same,

$$Z' = 2Z'.$$

In particular, since $(a + b)/2$ lies in Z', so do $a + b$ and $2a$. Suppose that $r > 1$ and that for $1 \leq j \leq r$ the element ja lies in Z'. If $(r + 1)a = 0$ then since p is prime the elements ja with $1 \leq j \leq r$ comprise all of the nonzero elements of $K(\delta)$ and so $h(t) \neq 0$ for $t \neq 0$ which contradicts our hypothesis. Therefore $(r + 1)a \neq 0$ and $ra \neq -a$, so $a + ra = (r + 1)a$ belongs to Z'. It follows that for all $r > 0$ the element ra lies in Z' and in particular $pa = 0$ lies in Z', which is a contradiction. Therefore, κ is of the form κ_t for some $t \neq 0$.

Remark (20.16):   The techniques used in the proof of Theorem 20.8 can be used to shorten the proof of Corollary 20.10 somewhat. Indeed, in the notation of that proof, we have

$$h(x + y).h(x - y) = h^2(x)G(y) - G(x)h^2(y).$$

We now claim that if $y \neq 0$ is such that $h(y) = 0$, then $G(y) = 0$. For, if not, then we conclude that $h(x) \neq 0$ implies that $h(x + y) \neq 0$, that is to say, the support of h is invariant under translation by y, and hence is the whole of $K(\delta)$ contradicting our assumption. From this and the above equation we see that if a, b with $a \neq \pm b$, belong to the support of h, and $(a + b)/2$ does not, then taking $x = (a + b)/2$, $y = (a - b)/2$ in the above equation, we get a contradiction since $h(x) = 0$ and hence $G(x)$ is also 0. Similarly we see that $(a - b)/2$ is also in its support. From here, the proof proceeds as above.

Corollary (20.17):   *The locus $\mathcal{L}$ has dimension ≤ 1.*

Proof: By Proposition 20.8, the locus $\mathcal{L}$ meets the hyperplane $E_x = 0$ in a finite set. Therefore $\mathcal{L}$ has dimension ≤ 1.

Our next task will be to compute the group of collineations of $\mathbf{P}(V^-)$ which leave the locus $\mathcal{L}$ invariant. That group will be denoted $Aut(\mathcal{L})$. We begin by showing that $PSL_2(\mathbf{F}_p)$ is isomorphic to a subgroup of $Aut(\mathcal{L})$, where $\mathbf{F}_p$ denotes the field with p elements. We identify the underlying sets of $\mathbf{F}_p$ and $K(\delta)$. If t is a nonzero element of $\mathbf{F}_p$ let A_t and B_t denote the operators on $V = V(\delta)$ given by

$$(A_t f)(x) = \zeta^{tx^2} f(x)$$

(20.18)

$$(B_t f)(x) = c \sum_{y \in \mathbf{F}_p} f(-y) \zeta^{txy}$$

for all f in V and all x in $\mathbf{F}_p$, where ζ is a primitive p-th root of unity and c is a certain nonzero constant which doesn't concern us[5] and which is independent of t. It is known ([W1],[A7], [K5],[D]) that there is one and only one representation ρ_t of $SL_2(\mathbf{F}_p)$ on V such that

$$\rho_t \left(\begin{pmatrix} 1 & 1 \\ 0 & 1 \end{pmatrix} \right) = A_t$$

(20.19)

$$\rho_t \left(\begin{pmatrix} 0 & 1 \\ -1 & 0 \end{pmatrix} \right) = B_t.$$

Furthermore, the center of $SL_2(\mathbf{F}_p)$ acts by a scalar on V^+ and on V^-, although the scalars are different values of ± 1. The action of $SL_2(\mathbf{F}_p)$ on V induces an action on $\bigotimes^2 V$. Call this representation ρ'_t. If we identify $\bigotimes^2 V$ with the space of all functions on $K(\delta,\delta)$, then for all γ in $SL_2(\mathbf{F}_p)$ we have

(20.20) $$(\rho'_t(\gamma) h)(x,y) = (\rho_t(\gamma) f)(x) \cdot (\rho_t(\gamma) g)(y)$$

where

$$h(x,y) = f(x) \cdot g(y).$$

It follows at once that the representation ρ'_t leaves Λ and Σ invariant.

Lemma (20.21):   The operator $\mathcal{T}$ on $\bigotimes^2 V$ which associates to $f(x,y)$ the function

(20.22) $$(\mathcal{T}f)(x,y) = f(x+y, x-y)$$

is an intertwining operator from ρ'_1 to ρ'_2.

Proof: We have to prove that if γ belongs to $SL_2(\mathbf{F}_p)$ then

(20.23) $$\mathcal{T} \circ (\rho'_1(\gamma)) = (\rho'_2(\gamma)) \circ \mathcal{T}.$$

It is enough to verify this on a set of generators of $\otimes^2 \mathcal{T}$. If f, g belong to V, let

$$h(x,y) = f(x) \cdot g(y).$$

Then we must show that

(20.24) $$\mathcal{T} \circ (\rho'_1(\gamma))(h) = (\rho'_2(\gamma)) \circ \mathcal{T}(h)$$

---

[5] For the precise value of this constant, see the paper [A3] of the first author, reproduced here as Appendix I of this book, or the published paper [Gé] of Gérardin.

or, what is the same, that

(20.25) $(\rho_1(\gamma)f)\, (x + y) \cdot (\rho_1(\gamma)g)\, (x - y) = (\rho_2'(\gamma)\circ T)\, (h)\, (x, y).$

To prove this, it is enough to verify it for

$$\gamma = \gamma_1 = \begin{pmatrix} 1 & 1 \\ 0 & 1 \end{pmatrix}$$

and for

$$\gamma = \gamma_2 = \begin{pmatrix} 0 & 1 \\ -1 & 0 \end{pmatrix},$$

since these elements generate $PSL_2(\mathbf{F}_p)$. For $\gamma = \gamma_1$ and any function $M(x, y)$ we have

(20.26) $(\rho_t'(\gamma)M)\, (x, y) = M(x, y)\, \zeta^{t(x^2 + y^2)}.$

In particular, we have

$$(T \circ \rho_1'(\gamma))(h)(x, y) \;=\; \rho_1'(\gamma)(h)\, (x + y, x - y)$$

$$\;=\; c\, h(x + y, x - y)\, \zeta^{(x+y)^2 + (x-y)^2}$$

(20.27)

$$\;=\; c\, h(x + y, x - y)\, \zeta^{2(x^2 + y^2)}$$

$$\;=\; (\rho_2'(\gamma)\circ T)(h)\, (x, y)$$

If $\gamma = \gamma_2$ then

(20.28) $(\rho_t'(\gamma)K)\, (x, y) = c \sum_{u, v \in \mathbf{F}_p} M(u, v)\, \zeta^{t(ux + vy)}$

for every M in $\otimes^2 V$. In particular, we see that

$$(T \circ \rho_1'(\gamma))(h)\, (x, y) \;=\; (\rho_1'(\gamma)(h))\, (x + y, x - y)$$

$$\;=\; c \sum_{u, v \in \mathbf{F}_p} h(u, v)\, \zeta^{u(x+y) + v(x-y)}$$

(20.29)

$$\;=\; c \sum_{u, v \in \mathbf{F}_p} h(u, v)\, \zeta^{(u+v)x + (u-v)y}$$

$$\;=\; c \sum_{a, b \in \mathbf{F}_p} h(a + b, a - b)\, \zeta^{2(ax + by)}$$

$$\;=\; (\rho_2'(\gamma)\circ T)\, (h)\, (x, y).$$

This completes the proof Lemma 20.12. We will refer to the operator T constructed in Lemma 20.12 as the **fundamental intertwining operator**.

Since $Gr \subset \mathbf{P}(\Lambda)$ is invariant under $GL(V^+)$, it is invariant under $\rho_2'(\gamma)$ for every γ in $SL_2(\mathbf{F}_p)$. Since T intertwines ρ_1' and ρ_2', it follows that when we identify Gr with a subvariety of $\mathbf{P}(\Sigma)$ via T, that subvariety of $\mathbf{P}(\Sigma)$ is invariant under $\rho_2'(\gamma)$ for every γ in $SL_2(\mathbf{F}_p)$. On the other hand, the variety V is invariant under $GL(V^-)$

and therefore under $\rho'_2(\gamma)$ for every γ in $SL_2(\mathbf{F}_p)$. Therefore the intersection $Gr \cap \mathcal{V}$ is invariant under $\rho'_2(\gamma)$ for every γ in $SL_2(\mathbf{F}_p)$. For any h in V, we have

$$(20.30) \qquad \rho'_1(\gamma)(h \vee h) = (\rho_2(\gamma)(h)) \vee (\rho_2(\gamma)(h))$$

for all γ in $SL_2(\mathbf{F}_p)$. Therefore $\mathcal{L}$, being the locus of all h in $\mathbf{P}(V^-)$ such that $h \vee h$ lies in $Gr \cap \mathcal{V}$, is invariant under $\rho_2(\gamma)$ for every γ in $SL_2(\mathbf{F}_p)$. Since the center of $SL_2(\mathbf{F}_p)$ acts trivially on $\mathbf{P}(V^-)$, it follows that the group of collineations of $\mathbf{P}(V^-)$ of the form $\rho_2(\gamma)$ is isomorphic to $PSL_2(\mathbf{F}_p)$. Since $\mathcal{L}$ contains the generators κ_t for $\mathbf{P}(V^-)$, it follows that the mapping of $PSL_2(\mathbf{F}_p)$ into $Aut(\mathcal{L})$ is injective. Therefore we can identify $PSL_2(\mathbf{F}_p)$ with a subgroup of $Aut(\mathcal{L})$. In characteristic 0, it is actually true that $PSL_2(\mathbf{F}_p)$ is equal to $Aut(\mathcal{L})$. To see this one can appeal to results of Brauer [Br] characterizing $PSL_2(\mathbf{F}_p)$ as a maximal finite subgroup of the collineation group of $\mathbf{P}(V^-)$. However, we can see directly that $PSL_2(\mathbf{F}_p) = Aut(\mathcal{L})$ as soon as we determine the 1-dimensional component of $\mathcal{L}$. If p is given, the same conclusion will hold in fields of sufficiently large characteristic (depending on p), but we do not know the precise exceptional set of primes. For the case $p = 11$, the first author has recently shown [A13] that the the prime 3 is exceptional. In fact, the automorphism group contains the Mathieu group M_{11} in this case!

That $\mathcal{L}$ has a one dimensional component follows from Theorem 20.30 below.

We will now show that if $\mathcal{L}$ contains a curve then the curve is irreducible. To see this, let $\mathcal{C}$ denote the union of all of the curves lying in $\mathcal{L}$. It will be enough to show that $\mathcal{C}$ is irreducible.

Lemma (20.31): *The tangent linear variety to $\mathcal{L}$ at the point κ_t is the line joining κ_t and κ_{3t}.*

Proof: Since $Aut(\mathcal{L})$ contains an automorphism which maps κ_1 to κ_t and κ_3 to κ_{3t}, we may assume that $t = 1$. If a cubic monomial $E_i E_j E_k$ with $1 \le i, j, k \le (p-1)/2$ does not vanish at κ_1, we must have $i = j = k = 1$. Therefore the first partial derivative of $\Phi_{w,x,y,z}$ at κ_1 will all vanish unless one of the terms of $\Phi_{w,x,y,z}$ is of the form $E_1^3 E_u$. Suppose that some partial derivative of of $\Phi_{w,x,y,z}$ does not vanish at κ_1. Then without loss of generality we may suppose that the first term of $\Phi_{w,x,y,z}$ is $E_1^3 E_u$. Furthermore, we may suppose that $w + x = w - x = y + z = 1$ and $y - z = u$. Then $w = 1$, $x = 0$, $u = 2y - 1 = 1 - 2z$ and $\Phi_{w,x,y,z}$ is equal to

$$(20.32) \qquad E_1^3 \cdot E_{2y-1} - E_{y+1} \cdot E_{y-1}^3 + E_y^3 \cdot E_{y-2}.$$

$\Phi_{w,x,y,z}$ will be nonzero if and only if y is not equal to 0, 1 or $(p+1)/2$. Furthermore, if $\Phi_{w,x,y,z}$ is nonzero then the first term of $\Phi_{w,x,y,z}$ will be the only term of the form $E_1^3 E_v$. Therefore the only partial derivative which will not vanish at κ_1 is the partial with respect to E_u. It follows that if κ is a point lying in the tangent linear variety to $\mathcal{L}$ at κ_1, we must have $E_u(\kappa) = 0$. If y is not equal to 0, 1 or $(p+1)/2$ then u is not equal to ± 1, 3 or 0. In particular we must have $E_u(\kappa) = 0$ for all $1 \le u \le (p+1)/2$ except $u = 1$ and $u = 3$. This shows that κ lies in the tangent linear variety to $\mathcal{L}$ at κ_1 if and only if κ lies on the line joining κ_1 and κ_3.

Corollary (20.33): *Assume that the curve part $\mathcal{C}$ of the locus $\mathcal{L}$ is nonempty.*

Then the curve C is irreducible. Furthermore, the point κ_t lies on C and is a simple point of C.

Proof: By Lemma 20.27, the tangent linear variety to $\mathcal{L}$ at κ_t has dimension 1. Since the curve part C of $\mathcal{L}$ is nonempty, the intersection of any component of C with the hyperplane $E_x = 0$ is nonempty. On the other hand, we know from Lemma 20.8 that the hyperplane $E_x = 0$ can only meet $\mathcal{L}$ in points of the form κ_t. Since C is contained in $\mathcal{L}$, it follows that $E_x = 0$ can only meet a component of C in points of the form κ_t. Therefore, every component of C contains a point of the form κ_u. By Lemma 20.27, we know that if κ_t lies on C then κ_t is a simple point of C. Consequently, only one irreducible of C can pass through each point κ_u. Since there are exactly $(p-1)/2$ points κ_u, it follows that the number of irreducible components of C is at most $(p-1)/2$. On the other hand, the orbit of κ_t under the action of $PSL_2(\mathbf{F}_p)$ contains all of the points of the form κ_u. Since C is invariant under $PSL_2(\mathbf{F}_p)$ and contains some point of the form κ_u, it must contain κ_t. In particular, κ_t is a simple point of C. If C_t is the component of C passing through κ_t and if γ is an element of $PSL_2(\mathbf{F}_p)$ mapping κ_t to κ_u, then γ must map the component C_t onto the component C_u passing through κ_u. If follows that $PSL_2(\mathbf{F}_p)$ acts transitively on the set of irreducible components of C. However, it is well known that $PSL_2(\mathbf{F}_p)$ has no nontrivial permutation representation of degree $< p$. Therefore C has only one irreducible component, that is to say, C is irreducible.

This is as far as we have been able to go using only the equations and without using in some way the relation of the locus $\mathcal{L}$ to moduli of elliptic curves. We now use that relation to establish the following theorem. We must emphasize, however, that the theorem is inferior to that of Vélu [Vél].

Theorem (20.34): Let $p \geq 7$ be a prime. Assume that the characteristic of k does not divide the order of $PSL_2(\mathbf{F}_p)$. The mapping ϕ given by $(E, L, x, y) \mapsto q_L$ maps the moduli space $\mathcal{M}_p$ of level p structures on an elliptic curve birationally onto the Zariski open subset C_0 of C obtained by removing the $PSL_2(\mathbf{F}_p)$-orbit of κ_1 from C.

Proof: Equation 19.6 can be rewritten in the form

(20.35) $\qquad q_L(x) \cdot q_L(y) = \begin{vmatrix} q_{L^2}\left(\dfrac{x+y}{2}\right) & q_{L^2}\left(\dfrac{x-y}{2}\right) \\ q_{L^2}\left(\dfrac{x+y}{2} + x_0\right) & q_{L^2}\left(\dfrac{x-y}{2} + x_0\right) \end{vmatrix}.$

If we define functions f, g, h on $K(p)$ by

$$f(x) \;=\; q_L(x)$$

(20.36) $\qquad g(x) \;=\; q_{L^2}(x)$

$$h(x) \;=\; q_{L^2}(x + x_0)$$

then we have $f \vee f = 2T(g \wedge h)$. Therefore by definition of $\mathcal{L}$ we conclude that $[f]$ lies in $\mathcal{L}$. The image $\mathcal{M}'_p$ of $\mathcal{M}_p$ in $\mathcal{L}$ is irreducible and the mapping of $\mathcal{M}_p$ onto its image is equivariant. Since there is no point of $\mathbf{P}(V^-)$ fixed by every element of $PSL_2(\mathbf{F}_p)$, the image $\mathcal{M}'_p$ cannot be a point. Therefore $\mathcal{M}'_p$ contains at least two points and since it must also be irreducible, it must actually be a curve. Since $\mathcal{C}$ is irreducible, it follows that $\mathcal{M}'_p$ is a dense open subset of $\mathcal{C}$. The completion of $\mathcal{M}_p$ is obtained by adding $(p^2 - 1)/2$ points which form a single $PSL_2(\mathbf{F}_p)$-orbit $\mathcal{O}$ of the completion, to $\mathcal{M}_p$. Denote that completion by $\overline{\mathcal{M}}_p$. Then the mapping ϕ extends to $\overline{\mathcal{M}}_p$ and must map $\overline{\mathcal{M}}_p$ onto all of $\mathcal{C}$. The image of $\mathcal{O}$ in $\mathbf{P}(V^-)$ must consist of points fixed by elements of order p, since the same is true of $\mathcal{O}$. But in $\mathbf{P}(V^-)$ the points fixed by elements of order p form a single orbit isomorphic to $\mathcal{O}$. Therefore, $\mathcal{M}'_p$ is obtained from $\mathcal{C}$ by removing a single $PSL_2(\mathbf{F}_p)$ orbit with $(p^2 - 1)/2$ elements, namely the orbit of κ_1. We will therefore be done if we can prove that the mapping of $\mathcal{M}_p$ onto $\mathcal{C}$ is birational. For that, it is enough to prove that $\mathcal{M}_p$ and $\mathcal{C}$ have the same genus. Let g denote the genus of $\mathcal{M}_p$ and let g' denote the genus of $\mathcal{C}$. Then we have

$$g - 1 = (p - 6) \cdot (p^2 - 1)/24$$

and $g \geq g'$. We know that there are points u, v of $\mathcal{M}_p$ such that the isotropy group of u in $PSL_2(\mathbf{F}_p)$ has 2 elements and the isotropy group of v in $PSL_2(\mathbf{F}_p)$ has exactly 3 elements. Therefore the isotropy group of $\phi(u)$ has order $2x$ for some integer $x \geq 1$ and the isotropy group of $\phi(v)$ has order $3y$ for some integer $y \geq 1$. The orbit of $\phi(u)$ has $(p^3 - p)/(4x)$ elements and the orbit of $\phi(v)$ has $(p^3 - p)/(6y)$ elements. Also, the isotropy group of κ_1 has p elements and the orbit of κ_1 has $(p^2 - 1)/2$ elements. Let $\mathcal{C}_0$ denote the quotient of $\mathcal{C}$ by the group $PSL_2(\mathbf{F}_p)$ and let g_0 denote the genus of $\mathcal{C}'$. Then we have

(20.37)
$$
\begin{aligned}
2 - 2g' &= \tfrac{1}{2}(p^3 - p) \cdot (2 - 2g_0) - \tfrac{1}{2}(p^2 - 1) \cdot (p - 1) \\
&\quad - \frac{1}{4x}(p^3 - p)(2x - 1) - \frac{1}{6y}(p^3 - p)(3y - 1).
\end{aligned}
$$

Since $g_0 \geq 0$ it follows that

(20.38)
$$
\begin{aligned}
2 - 2g' &\leq (p^3 - p) - \tfrac{1}{2}(p^2 - 1) \cdot (p - 1) - (p^3 - p) \\
&\qquad + \frac{p^3 - p}{4x} + \frac{p^3 - p}{6y} \\
&= \tfrac{1}{2}(p^3 - p) \cdot [\frac{1}{2x} + \frac{1}{3y} + \frac{1}{p} - 1] \\
&\leq \tfrac{1}{2}(p^3 - p)[\frac{1}{2} + \frac{1}{3} + \frac{1}{p} - 1] \\
&= \tfrac{1}{2}(p^3 - p) \cdot \frac{[3p + 2p + 6 - 6p]}{6p} \\
&= -\tfrac{1}{12}(p - 6) \cdot (p^2 - 1).
\end{aligned}
$$

Therefore

$$(20.39) \qquad g' - 1 \geq \frac{1}{24}(p - 6) \cdot (p^2 - 1),$$

so $g' \geq g$ and we are done.

The following lemma is undoubtedly well known but since we do not know a suitable reference, we include a proof.

Lemma (20.40): Let $p \geq 7$ be a prime. In characteristic 0 or sufficiently large characteristic (depending on p) the automorphism group of $\mathcal{M}_p$ is precisely $PSL_2(\mathbf{F}_p)$.

Proof: First we will prove the result in characteristic 0 and for that we may assume we are working over the field of complex numbers. Let $\mathcal{G}$ denote the automorphism group of $\mathcal{M}_p$. The genus of $\mathcal{M}_p$ is

$$1 + \frac{(p^2 - 1)(p - 6)}{24}.$$

By a theorem of Hurwitz, the order $|\mathcal{G}|$ of $\mathcal{G}$ cannot exceed $84(g-1)$, where g is the genus of $\mathcal{M}_p$. Therefore, we have the following inequality for the index of $PSL_2(\mathbf{F}_p)$ in $\mathcal{G}$:

$$\frac{|\mathcal{G}|}{|PSL_2(\mathbf{F}_p)|} \leq \frac{84(g - 1)}{|PSL_2(\mathbf{F}_p)|} = \frac{14}{3}\frac{p - 6}{p} < 5,$$

hence the index is at most 4. Suppose that $\mathcal{G}$ does not equal $PSL_2(\mathbf{F}_p)$. Consider the permutation representation ϖ of $\mathcal{H}$ on cosets of $PSL_2(\mathbf{F}_p)$. The kernel $ker(\varpi)$ of ϖ is the largest normal subgroup[6] of $\mathcal{G}$ contained in $PSL_2(\mathbf{F}_p)$. Since there are at most 4 cosets, the index of $ker(\varpi)$ in H is at most $\leq 4! = 24$. Since $PSL_2(\mathbf{F}_p)$ is a simple group of order > 24, the kernel must contain $PSL_2(\mathbf{F}_p)$, so $PSL_2(\mathbf{F}_p)$ is a normal subgroup of $\mathcal{G}$. Now let γ be an element of $\mathcal{G}$ not in $PSL_2(\mathbf{F}_p)$. Since γ normalizes $PSL_2(\mathbf{F}_p)$, it induces an automorphism $\bar{\gamma}$ of the orbit space $\mathbf{P}^1$ for $PSL_2(\mathbf{F}_p)$ acting on $\mathcal{M}_p$. Since there is exactly one orbit for each of the branch points of order 2,3 and p, it follows that the corresponding points of $\mathbf{P}^1$ are fixed by $\bar{\gamma}$. Therefore, $\bar{\gamma}$ acts trivially on $\mathbf{P}^1$, since it fixes three points. It now follows from Galois theory that $\mathcal{G}$ equals $PSL_2(\mathbf{F}_p)$. This proves the result in characteristic 0. If, for given p, the result is not true in characteristic q for sufficiently large primes q, then let (q_n) be a strictly increasing sequence of primes such that the result is false in characteristic q_n and let $\mathcal{F}_n$ be an algebraically closed field of characteristic q_n in which one has an automorphism γ_n of $\mathcal{M}_p$ not lying in $PSL_2(\mathbf{F}_p)$. We can embed $\mathcal{M}_p$ in a projective space using a suitable power of the canonical line bundle on $\mathcal{M}_p$. Having done so, the automorphism group of $\mathcal{M}_p$ is faithfully represented as a group of collineations leaving the image of the curve $\mathcal{M}_p$ invariant. Therefore, we may suppose that γ_n is concretely described by a matrix γ_n' with entries in $\mathcal{F}_n$, the size of the matrix being independent of n. Let $\mathcal{F}$ denote the field obtained by taking the cartesian product of all of the fields $\mathcal{F}_n$ and modding out by a maximal ideal containing the direct sum of all of the fields $\mathcal{F}_n$. One can then verify that $\mathcal{F}$ is an

---

[6] This is given as an exercise in Herstein's *Topics in Algebra*.

algebraically closed field[7] of characteristic 0. The family of matrices γ'_n determines a matrix of the same size with entries in $\mathcal{F}$ leaving invariant the modular curve $\mathcal{M}_p$ over $\mathcal{F}$. That automorphism will not lie in $PSL_2(\mathbf{F}_p)$, contradicting what we proved in the case of characteristic 0.

Corollary (20.41):   *If the characteristic of k is sufficiently large or is zero, then*

$$Aut(\mathcal{L}) = PSL_2(\mathbf{F}_p).$$

Proof: The morphism $\mathcal{M}_p \to \mathcal{C}$ is a desingularization, hence a normalization. Since normalization is a functor, every automorphism of $\mathcal{C}$ induces an automorphism of $\mathcal{M}_p$. But in characteristic 0 or sufficiently large characteristic (depending on p), by Lemma 20.40, the automorphism group of $\mathcal{M}_p$ is precisely $PSL_2(\mathbf{F}_p)$, so

$$Aut(\mathcal{C}) = PSL_2(\mathbf{F}_p).$$

Since $\mathcal{C}$ spans $\mathbf{P}(V^-)$ and since $Aut(\mathcal{L})$ leaves $\mathcal{C}$ invariant, we conclude that

$$Aut(\mathcal{L}) \subseteq Aut(\mathcal{C}) = PSL_2(\mathbf{F}_p).$$

Since we have proved the opposite inclusion earlier, we are done.

We are unable to show by our elementary methods that $\mathcal{C}$ is nonsingular or that the locus $\mathcal{L}$ contains no isolated points. The equations 19.13 were obtained independently by Vélu who investigated the universal elliptic curve in his thesis [Vé 1]. He was able to use the family of elliptic curves over $\mathcal{C}$ to prove that $\mathcal{C}$ is nonsingular and has no isolated points. He was apparently unaware that the equations 19.13 were actually discovered by Klein [K1-4], [K-F], who also computed the degree of $\mathcal{C}$ using automorphic functions and made a close study of some special cases. Klein also gave the equations defining the universal family of elliptic curves. Even after a century, Klein's accomplishments are remarkable to behold.

§21   Level 3 Structure and Invariants of Symplectic Groups

In this section, we will examine the equations introduced in the discussion of Case 2 of §19. We begin with a simple and instructive example. Recall that the characteristic doesn't divide the order of $K(\delta)$.

Let us suppose that $n = 2$ and $\delta = (3,3)$. Let

$$(21.1) \qquad \begin{aligned} Y_0 &= q_L(0,0) \\ Y_1 &= q_L(0,1) \\ Y_2 &= q_L(1,0) \\ Y_3 &= q_L(1,1) \\ Y_4 &= q_L(1,2). \end{aligned}$$

Then the point $[Y_0, Y_1, Y_2, Y_3, Y_4]$ of $\mathbf{P}^4$ lies in the Hessian of the quartic hypersurface

---

[7]   If we take all the $\mathcal{F}_n$ to be countable, as we may, then $\mathcal{F}$ will be isomorphic to the field of complex numbers.

defined by the equation

(21.2) $$Y_0^4 + 8Y_0 \cdot (Y_1^3 + Y_2^3 + Y_3^3 + Y_4^3) + 48\,Y_1Y_2Y_3Y_4 = 0.$$

The Hessian is defined by the determinant

(21.3)
$$\begin{vmatrix} Y_0^2 & Y_1^2 & Y_2^2 & Y_3^2 & Y_4^2 \\ Y_1^2 & Y_0Y_1 & Y_3Y_4 & Y_2Y_4 & Y_2Y_3 \\ Y_2^2 & Y_3Y_4 & Y_0Y_2 & Y_1Y_4 & Y_1Y_3 \\ Y_3^2 & Y_2Y_4 & Y_1Y_4 & Y_0Y_3 & Y_1Y_2 \\ Y_4^2 & Y_2Y_3 & Y_1Y_3 & Y_1Y_2 & Y_0Y_4 \end{vmatrix} = 0.$$

Referring to the definition of the matrix S in 19.22, one verifies very easily that this determinant is the 5×5 minor of S consisting of the (x, y)-th entries of S with x and y running over the points $(0,0)$, $(0,1)$, $(1,0)$, $(1,1)$ and $(1,2)$ of $K(3,3)$. By Proposition 19.25, the rank of S is $\leq 2^2 = 4$, so this 5×5 minor must vanish.

Remark (21.4): This result is due to Burkhardt [Bu] who proved it by a different method. The quartic polynomial is an invariant of the finite symplectic group[8] $Sp(\mathbf{F}_3^2)$ acting on the space of even functions on $\mathbf{F}_3^2$. This polynomial and its Hessian appear in Burkhardt's paper as the invariants J_4 and J_{10} on page 208. We will now generalize them to the case of higher dimensional abelian varieties with a symmetric invertible sheaf of level $(3, 3, \ldots, 3)$.

Let X be an abelian variety of dimension g and let L be an ample symmetric invertible sheaf on X with trivial Arf invariant and whose level is $\delta = (3, 3, \ldots, 3)$. For every element u of $K(\delta)$, let

$$Y_u = q_L(u).$$

Then by Lemma 18.10, we have

$$Y_u = Y_{-u}.$$

Let $\mathcal{O}$ be a full set of representatives for $K(\delta)$ modulo the equivalence relation defined by $u = \pm v$. Let

$$S_0 : \mathcal{O} \times \mathcal{O} \to k$$

denote the matrix whose rows and columns are indexed by the elements of $\mathcal{O}$ and whose (u, v)-th entry is given by

$$S_0(u, v) = Y_{u+v}Y_{u-v}.$$

Then S_0 is contained in the matrix S of 19.22 and the rank of S_0 therefore does not exceed 2^g. It is convenient to introduce the strict lexicographic ordering $\prec$ on $\mathcal{O}$. If we write $x \prec y$, we will imply that $x \neq y$. This defines a total ordering of $\mathcal{O}$.

We will denote by F_g the quartic polynomial defined by

(21.5) $$F_g = Y_0^4 + 8 \cdot Y_0 \cdot \left(\sum_{0 \prec u} Y_u^3\right) + 8 \cdot \sum_{0 \prec u \prec v} Y_uY_vY_{u+v}Y_{u-v}.$$

---

[8] We use the notation of Weil's paper [W1] to denote symplectic groups. In this case the group is more commonly denoted $Sp_4(\mathbf{F}_3)$.

Note that for $g = 2$, the polynomial F_g agrees with Burkhardt's quartic as given in equation 21.2. For u, v belonging to $\mathcal{O}$, we can compute the second partial derivative of F_g with respect to Y_u and Y_v and we obtain

(21.6) $$\frac{\partial^2 F_n}{\partial Y_u \partial Y_v} = \begin{cases} 12 \cdot Y_0^2 & \text{if } u = v = 0 \\ 48 \cdot Y_0 Y_u & \text{if } u = v \neq 0 \\ 24 \cdot Y_v^2 & \text{if } u = 0 \prec v \\ 48 \cdot Y_{u+v} Y_{u-v} & \text{if } 0 \prec u \prec v \end{cases}$$

Denote by C the matrix of second partials of F_g. Let

$$P : \mathcal{O} \times \mathcal{O} \to k$$

denote the diagonal $N \times N$ matrix, with $N = (3^g + 1)/2$, whose rows and columns are indexed by the elements of $\mathcal{O}$ and whose diagonal entries p_u are equal to 2 for $u \neq 0$ and to 1 for $u = 0$. Let $Q = 12 \cdot P$. Then it is easy to see that $P \cdot S_0 \cdot Q$ is equal to the matrix of second partial derivatives of F_g. Since P and Q are both nonsingular, it follows that the rank of S_0 is equal to the rank of the matrix of second partials of F_g. We have therefore proved the following result.

Theorem (21.7): *Let X be an abelian variety of dimension g and let L be an ample symmetric invertible sheaf on X with trivial Arf invariant and level*

$$\delta = (3, 3, \ldots, 3).$$

Then the rank of the matrix of second partial derivatives of F_g at the point

$$Y = [q_L(u); \, u \in \mathcal{O}]$$

is $\leq 2^g$, where $\mathcal{O}$ is a full set of representatives for the equivalence relation

$$x = \pm y$$

on $K(\delta)$. In particular, the theta constants q_L map the moduli space of abelian varieties with ample totally symmetric line bundle of level $(3, \ldots, 3)$ into the rank $\leq 2^g$ locus of the Hessian of the quartic $F_g = 0$.

Our next goal is to prove that if $g > 1$ then F_g is the unique quartic invariant for the Weil representation of $Sp(\mathbf{F}_3^g)$ on the space of even k-valued functions on $\mathbf{F}_3^g$. This will be accomplished in Theorem 21.35. We begin with several pages of preliminary arguments in which we establish the existence and uniqueness of a quartic invariant. Let us recall the explicit transformations [W1] which define the projective representation of $Sp(\mathbf{F}_3^g)$ on the projective space $\mathbf{P}(V(\delta))$. We follow the notation of [W1]. For A in $GL_g(\mathbf{F}_3)$, the matrix

(21.8) $$d_0(A) = \begin{pmatrix} A & 0 \\ 0 & {}^t A^{-1} \end{pmatrix} \in Sp(\mathbf{F}_3^g)$$

acts on $\mathbf{P}(V(\delta))$ by

(21.9) $$(\mathbf{d}_0(A)\Phi)(x) = \Phi(x \cdot A)$$

If B is a symmetric $g \times g$ matrix with entries in $\mathbf{F}_3$, the matrix

$$(21.10) \qquad (t_0(B)\Phi)(x) = \begin{pmatrix} I & B \\ 0 & I \end{pmatrix} \in Sp(\mathbf{F}_3^g)$$

acts on $\mathbf{P}(V(\delta))$ by

$$(21.11) \qquad (t_0(B)\Phi)(x) = \epsilon^{-B[x]}\,\Phi(x)$$

where ϵ is a primitive cube root of unity. Finally, the matrix

$$(21.12) \qquad d_0'(I_g) = \begin{pmatrix} 0 & I_g \\ -I_g & 0 \end{pmatrix},$$

where I_g is a $g \times g$ identity matrix, acts on $\mathbf{P}(V(\delta))$ by

$$(21.13) \qquad (d_0'(I_g))(x) = \frac{1}{\sqrt{3^g}} \sum_{y \in \mathbf{F}_3^g} \Phi(-y)\,\epsilon^{-x \cdot y}.$$

The Weil representation depends on the choice of ϵ. It is well known that the Weil representation of $Sp(\mathbf{F}_3^g)$ lifts to a linear representation (a proof can be found in [A10]). Explicit transformations defining that linear representation may be found in [A10] and in [Gé]. However, we will only need to know that equation 21.11 also defines the lifting of the action of $t_0(B)$ and equation 21.9 defines the lifting of the action of $d_0(A)$ in case A has determinant 1. To see why it is true in this case, let

$$W^+ = \{f : K(\delta) \to k \mid (\forall u)(f(u) = f(-u))\}$$

denote the subspace of $V(\delta)$ consisting of even function and let

$$W^- = \{f : K(\delta) \to k \mid (\forall u)(f(u) = -f(-u))\}$$

denote the space of odd functions. Both W^+ and W^- are invariant under the Weil representation of $Sp(\mathbf{F}_3^g)$. Since the kernel of the homomorphism

$$(21.14) \qquad SL(W^\pm) \to PSL(W^\pm)$$

is a cyclic group whose order divides $(3^g \pm 1)/2$, its order is not divisible by 3. Since the right hand side of 21.11 evidently has order 3, it must be the lifting of $t_0(B)$. Similarly, if A has order 3 then the right hand side of 21.8 defines the lifting of $d_0(A)$. Since the elements of order 3 in $GL_g(\mathbf{F}_3)$ generate the subgroup $SL_g(\mathbf{F}_3)$, the right hand side of 21.8 defines the lifting of $d_0(A)$ for any A of determinant 1.

Suppose F is a homogeneous quartic polynomial on W^+ which defines a quartic hypersurface in $\mathbf{P}(W^+)$ invariant under the Weil representation of $Sp(\mathbf{F}_3^g)$. Write

$$(21.15) \qquad F(t) = \sum c_{wxyz}\, Y_w Y_x Y_y Y_z.$$

Then $t_0(B)$ transforms F into

$$(21.16) \qquad \sum c_{wxyz}\, Y_w Y_x Y_y Y_z\, \epsilon^{-B[w]-B[x]-B[y]-B[z]}.$$

Since the right hand side of 21.11 defines the lifting of the Weil representation on the element $t_0(B)$, the right hand side of 21.15 must equal 21.16. Therefore

(21.17) $c_{wxyz} = 0$

unless

(21.18) $B[w] + B[x] + B[y] + B[z] = 0.$

for every $B = {}^t B$ in $M_g(\mathbf{F}_3)$.

Lemma (21.19):   *Let w, x, y, z lie in $\mathbf{F}_3^g$ and suppose that for every quadratic form ϕ on $\mathbf{F}_3^g$ we have*

(21.20) $\phi(w) + \phi(x) + \phi(y) + \phi(z) = 0.$

Then

(21.21) $\begin{aligned} \pm y &= w \pm x \\ \pm z &= w \mp x \end{aligned}$

where the signs of w and z are chosen independently and where the signs of x in the two equations are opposite but otherwise independent of the signs of w and z.

Proof: If $g \leq 1$, this is very easy to verify. We first observe that (21.20) implies that any linear l form which vanishes on two of w, x, y, z vanishes on all of them. Indeed, if we take ϕ to be l^2, we obtain

$$l(y)^2 + l(z)^2 = 0$$

and since the left hand side is the sum of two numbers which are either 0 or 1, we conclude that both must be 0. From this observation, we conclude that the span of any two elements is the same as the span of all four. If this span has dimension ≤ 1, we are reduced to the case $g \leq 1$. Therefore, we may suppose that the span of any two of w, x, y, z has dimension 2. Then y, z belong to the space spanned by w, x and cannot be 0, $\pm w$ or $\pm x$. The only remaining elements of the span of w, x are $\pm(w + x)$ and $\pm(w - x)$. Since y, z are linearly independent, it follows that one of them must be of the form $\pm(w + x)$ and the other of the form $\pm(w - x)$. This proves the lemma.

The following corollary is an immediate consequence of the lemma.

Corollary (21.22):   *If F is a quartic form on W^+ which defines a hypersurface in $\mathbf{P}(W^+)$ invariant under the action of $Sp(\mathbf{F}_3^g)$, then F is of the form*

(21.23) $\sum_{u,v \in \mathbf{F}_3^n} d_{uv} Y_u Y_v Y_{u+v} Y_{u-v}$

where, referring to our notation in 21.15 we have

$$d_{uv} = c_{u,v,u+v,u-v}.$$

We can rewrite 21.39 in the form

$$(21.24) \qquad aY_0^4 + Y_0 \cdot \sum_{u \neq 0} b_u Y_u^3 + \sum_{0 \neq u \neq \pm v \neq 0} d_{uv} Y_u Y_v Y_{u+v} Y_{u-v}.$$

In other words, the last summation runs over all u, v which are linearly independent, the middle summation runs over all u, v with 1-dimensional span and the first term represents those u, v with 0-dimensional span. For every one dimensional subspace of $\mathbf{F}_3^g$, let

$$(21.25) \qquad\qquad\qquad Y_\lambda = Y_u$$

where u is any nonzero element of λ. We can then rewrite 21.40 more conveniently in the form

$$(21.26) \qquad aY_0^4 + Y_0 \cdot \sum_\lambda \beta_\lambda Y_\lambda^3 + \sum_\pi \delta_\pi \prod_{\lambda \subset \pi} Y_\lambda$$

where the β_λ and δ_π are coefficients, where the middle summation runs over all subspaces λ of $\mathbf{F}_3^g$ of dimension 1 and where the last summation runs over all subspaces π of dimension 2 and where the product runs over all lines λ in the plane π. Since the group $SL_g(\mathbf{F}_3)$ acts transitively on the subspaces of a given dimension of $\mathbf{F}_3^g$ and since elements of that group act, according to the remarks following 21.13, by the right hand side of 21.9, it follows that the coefficients β_λ are independent of λ and the coefficients δ_π are independent of π. It follows that we can write 21.42 more simply as

$$(21.27) \qquad a \cdot Y_0^4 + b \cdot Y_0 \cdot \sum_\lambda Y_\lambda^3 + c \cdot \sum_\pi \prod_{\lambda \subset \pi} Y_\lambda.$$

If the quartic invariant F were not unique up to a constant multiple, we could find another quartic invariant

$$(21.28) \qquad F' = a'Y_0^4 + b'Y_0 \cdot \sum_\lambda Y_\lambda^3 + c' \sum_\pi \prod_{\lambda \subset \pi} Y_\lambda$$

not proportional to F. Then by taking a suitable linear combination of F and F' we could eliminate the summation over 2-planes π and obtain a quartic invariant of the form

$$(21.29) \qquad a'' \cdot Y_0^4 + b'' \cdot Y_0 \cdot \sum_\lambda Y_\lambda^3,$$

which is reducible. The group $Sp(\mathbf{F}_3^g)$ must permute the irreducible components of an invariant hypersurface and since $Sp(\mathbf{F}_3^g)/\{\pm 1\}$ is a simple group of order $> 4!$, every factor of F'' must be invariant. But Y_0 is a factor which is clearly not invariant under $d_0'(I_g)$. We have therefore proved that **there is, up to a constant multiple, at most one quartic invariant for the action of** $Sp(\mathbf{F}_3^g)$ **on** $\mathbf{W}^+$.

To see that a quartic invariant for $Sp(\mathbf{F}_3^g)$ actually exists, recall that the Weil representation depended on a choice for a cube root of unity. Let ϵ be one such choice and let ρ denote the corresponding Weil representation. Let ρ' denote the

Weil representation one obtains if ϵ is replaced by ϵ^2. Then ρ' is the dual of ρ and $Sym^2(\rho')$ is the dual of $Sym^2(\rho)$. On the other hand, the representations ρ, ρ' are modelled on the same space $V(\delta)$. The space W^+ of even functions on $K(\delta)$ is invariant under the Weil representations of $Sp(\mathbf{F}_3^g)$ and the space $Sym^2(W^+)$ may be identified with the space of all functions $f(u,v)$ on $K(\delta) \times K(\delta)$ such that

$$(21.30) \qquad\qquad f(u,v) = f(v,u) = f(-u,v)$$

for all u, v in $K(\delta)$. Define the operator $\mathcal{T}$ on $V(\delta,\delta)$ as in 19.8 by the rule

$$(21.31) \qquad (\mathcal{T}f)(u,v) = f\left(\frac{u+v}{2}, \frac{u-v}{2}\right) = f(-u-v, -u+v).$$

Then the operator $\mathcal{T}$ leaves the space $Sym^2(W^+)$ invariant and intertwines $Sym^2(\rho)$ with $Sym^2(\rho')$; the proof of this assertion is quite similar to the proof of Lemma 20.17 and will be left to the reader. We can therefore view $\mathcal{T}$ as a $Sp(\mathbf{F}_3^g)$-equivariant isomorphism of $Sym^2(W^+)$ onto its dual space and therefore as a linear form $\$$ on $Sym^2(W^+) \otimes Sym^2(W^+)$. We can decompose this tensor product as the sum of $Sym^4(W^+)$ and other components in an equivariant manner and we can consider the restriction of $\$$ to an invariant linear functional $\$_0$ on $Sym^4(W^+)$. Then $\$_0$ is the quartic invariant which we are seeking and our task is simply to show that $\$_0$ is not zero. For that it will be enough to show that for some f in $Sym^2(W^+)$ we have

$$\$(f \otimes f) \neq 0$$

or, what is the same, that

$$(\mathcal{T}f)(f) \neq 0.$$

The linear form

$$\beta : Sym^2(W^+) \otimes Sym^2(W^+) \to k$$

given by

$$(21.32) \qquad \beta(f, f') = \sum_{w,z \in K(\delta)} f(w,z) \cdot f'(w,z)$$

is a nondegenerate pairing which one can show is invariant with respect to the action $Sym^2(\rho) \otimes Sym^2(\rho')$ of $Sp(\mathbf{F}_3^g)$ on $Sym^2(W^+) \otimes Sym^2(W^+)$. Therefore we only have to check that for some f in $Sym^2(W^+)$ of the form $h \otimes h$ with $h \in W^+$ we have

$$(21.33) \qquad \sum_{w,z \in K(\delta)} f(w,z) \cdot f(-w-z, -w+z) \neq 0.$$

But if we take $f(w,z)$ to be the function which is 1 at $(0,0)$ and 0 elsewhere on $K(\delta,\delta)$ then f lies in $Sym^2(W^+)$ and we have

$$(21.34) \qquad \sum_{w,z \in K(\delta)} f(w,z) \cdot f(-w-z, -w+z) = 1.$$

This proves that the quartic invariant exists. We now have the following result.

Theorem (21.35): Let g be an integer greater than 1 and let ρ_+ denote the restriction to W^+ of the Weil representation ρ of $Sp(\mathbf{F}_3^g)$. Then there is one and, up to a scalar factor, only one invariant polynomial of degree 4 for ρ. It is given explicitly by equation 21.5.

Proof: Everything has been proved except for the assertion that the invariant is equal to 21.5. To see that, we use the expression 21.32 for the bilinear pairing β on $Sym^2(W^+)$. For all u in $K(\delta)$, denote by ϕ_u the function on $K(\delta)$ given by

$$(21.36) \qquad\qquad \phi_u(z) = \begin{cases} 1 & \text{if } z = u \\ 0 & \text{if not.} \end{cases}$$

Then the functions $\phi_u + \phi_{-u}$ with u in $\mathcal{O}$ form a basis for W^+. Let

$$(21.37) \qquad\qquad Y = \sum_{u \in K(\delta)} Y_u \phi_u$$

We then have

$$(21.38) \qquad\qquad Y \otimes Y = \sum_{u,v \in K(\delta)} Y_u Y_v\, \phi_u \otimes \phi_v.$$

Let

$$(21.39) \qquad\qquad f(Y) = \beta(Y \otimes Y, Y \otimes Y).$$

Then f is a quartic invariant and we have

$$f(Y) = \sum_{w,y \in K(\delta)} (Y \otimes Y)(w,y) \cdot (Y \otimes Y)(-w - y, -w + y)$$

$$(21.40)$$
$$= \sum_{w,y \in K(\delta)} \Big(\sum_{u,v \in K(\delta)} Y_u Y_v \cdot (\phi_u \otimes \phi_v)(w,y) \Big)$$
$$\cdot \Big(\sum_{s,t \in K(\delta)} Y_s Y_t \cdot (\phi_s \otimes \phi_t)(-w - y, -w + y) \Big)$$

$$= \sum_{w,y \in K(\delta)} Y_w Y_y Y_{-w-y} Y_{-w+y}.$$

Since the expression

$$Y_w Y_y Y_{-w-y} Y_{-w+y}$$

depends only on the span of w and y, we can collect terms according to the span of w and y and we obtain

$$(21.41) \qquad\qquad f(Y) = Y_0^4 + 8 Y_0 \sum_\lambda Y_\lambda^3 + 48 \sum_\pi \prod_{\lambda \subset \pi} Y_\pi$$

where the middle summation runs over all subspaces λ of $\mathbf{F}_3^g$ of dimension 1 and the last summation over all subspaces π of dimension 2. The middle summation of 21.41 evidently equals the corresponding summation of 21.5. As for the last term, if π is any 2-dimensional subspace of $\mathbf{F}_3^g$, there are exactly six ways in which we can choose u, v in $\mathcal{O} \cap \pi$ such that

$$0 \prec u \prec v.$$

Therefore, the last summation of 21.41 equals the last summation of 21.5, and we are done.

Chapter V

Invariant Theory, Arithmetic and Vector Bundles

§22   The Mysterious Role of Invariant Theory

In §20 we saw how the group of symmetries of the modular curve helped to make the study of the equations 19.13 more tractable from our elementary point of view. In §21, an invariant of degree 4 for the symplectic group dominated the study of the moduli space of theta structures of level $\delta = (3, 3, \ldots, 3)$ on abelian varieties. In this section, we discuss more examples of this phenomenon and try to formulate a general perspective.

Example (22.1):   At the end of §20, we gave the equation of the modular curve of level 7 as

(22.2) $$X^3 Y + Y^3 Z + Z^3 X = 0$$

in $\mathbf{P}^2$. The left hand side is the unique quartic invariant[9] for $PSL_2(\mathbf{F}_7)$ in this representation.

---

[9] This quartic was discovered by Felix Klein. Klein went further and determined the full ring of invariants for a 3 dimensional irreducible complex representation of $PSL_2(\mathbf{F}_7)$. For details see [K1] and [K-F]. One can give simple proofs that this quartic is the unique quartic invariant for $PSL_2(\mathbf{F}_7)$. First, a character computation shows that there exists a unique quartic invariant for this group. Either from the character or from the explicit description of the representation as the odd part of the Weil representation, one knows that with respect to a suitable basis, the upper triangular subgroup of order 21 acts in the following way: it maps to the group generated by the diagonal matrix with entries $[\zeta, \zeta^4, \zeta^2]$, where ζ is a primitive 7th root of unity, and by the cyclic permutation of order 3 of the coordinates. The element of order 7 has for its invariant quartics the linear combinations of the 3 monomials $X^3 Y$, $Y^3 Z$ and $Z^3 X$. It is then obvious that Klein's quartic is, up to a constant multiple, the only linear combination of these 3 which is invariant under the cyclic permutation. A similar argument shows that (22.4) is the unique cubic invariant for $PSL_2(\mathbf{F}_{11})$ acting on $\mathbf{C}^5$.

Example (22.3):   The following matrix

$$\begin{pmatrix} w & v & 0 & 0 & z \\ v & x & w & 0 & 0 \\ 0 & w & y & x & 0 \\ 0 & 0 & x & z & y \\ z & 0 & 0 & y & v \end{pmatrix}$$

is, up to a factor of 2, the matrix of second partial derivatives of a cubic invariant for $PSL_2(\mathbf{F}_{11})$, namely

(22.4) $$v^2 w + w^2 x + x^2 y + y^2 z + z^2 v = 0.$$

The modular curve is then the singular locus of the Hessian of this cubic threefold.

This result was discovered by Felix Klein [K2], [K-F]. Closer examination of his work shows that he actually did not distinguish carefully between the rank 3 locus of the matrix of second partials and the singular locus of the Hessian as a whole. That they are in fact the same is shown in [A2], [A16].

Example (22.5):   A Kummer surface is the image in $\mathbf{P}^3$ of an abelian surface under the linear system $\Gamma(L)$ where L is an invertible sheaf of level $\delta = (2, 2)$. The Heisenberg group $G(L)$ acts on $\mathbf{P}^3$ and the Kummer surface is a quartic invariant.

Example (22.6):   Let E be an elliptic normal curve in $\mathbf{P}^4$. Then E has degree 5 and is invariant under the Heisenberg group $G(5)$. We know from the Riemann-Roch theorem that E lies on 5 quadrics and in fact these quadrics can easily be shown to define E. Let $\phi_0, \phi_1, \phi_2, \phi_3, \phi_4$ be a basis for the quadrics containing E. If $x = [x_0, x_1, x_2, x_3, x_4]$ is a point of $\mathbf{P}^4$, denote by Q_x the quadric

(22.7) $$\sum_{i=0}^{4} x_i \phi_i = 0$$

in $\mathbf{P}^4$. Let J denote the quintic hypersurface in $\mathbf{P}^4$ consisting of all x such that Q_x is singular. Then J is an invariant for the action of the Heisenberg group. In fact, J is the chord locus of the elliptic normal curve E and the singular locus of J is E. The first author believes he found this example in a work of Halphen, but for some reason has been unable to find it again.

Example (22.8):   In their joint paper [N-R 1], Narasimhan and Ramanan have the following example. Let J denote the Jacobian variety of a curve of genus 3 which is not hyperelliptic. We can map J into $\mathbf{P}^7$ using twice the theta divisor. The image is the Kummer variety considered by J. Coble [Co2], who suggested that the Kummer variety might be the singular locus of a certain quartic invariant for the Heisenberg group $G(2, 2, 2)$. Narasimhan and Ramanan were able to prove this using their results on moduli of rank 2 vector bundles.

Example (22.9):   Let X be the Jacobian variety of a curve $\mathcal{C}$ of genus 2 and let $L = \mathcal{O}_X(\mathcal{C})$. Consider the of X into $\mathbf{P}^8$ determined by L^3. By a theorem of Lefschetz, we know that this mapping is an embedding. We will therefore identify X with its image. The locus X is invariant under the Heisenberg group $G(3, 3)$ and one can show, using the Riemann-Roch theorem that X lies in an eight dimensional linear system of quadrics which form a $G(3,3)$ invariant space of quadrics. One can show that the partial linear system of quadrics through X is the system of polar quadrics of a cubic invariant for $G(3,3)$. Denote that cubic by $\mathcal{K}$. Then X must lie on the singular locus of $\mathcal{K}$. Dolgachev has kindly brought to our attention the fact that in the paper [Co1], Coble proved that X is actually equal to the singular locus of $\mathcal{K}$. David Grant [Gra] wrote down the cubic explicitly.

Coble also proved the remarkable result that Burkhardt's quartic hypersurface is its own Steinerian.[10] Apart from the intrinsic interest of such a quartic, this implies that Burkhardt's quartic hypersurface is birationally equivalent to its Hessian in a natural way. As we have noted at the beginning of §21, the Hessian of Burkhardt's quartic is defined by the relations among the 5 distinct theta constants of level $(3,3)$ and trivial Arf invariant. If one instead considers theta constants of level $(3,3)$ and nontrivial Arf invariant, one obtains 4 distinct theta constants and the addition formula implies no relations among them. Hence the abelian surfaces with ample symmetric invertible sheaves of level $(3,3)$ and nontrivial Arf invariant correspond to points of $\mathbf{P}^3$, the four homogeneous coordinates being the four odd theta constants. As Burkhardt has already observed, this $\mathbf{P}^3$ can be mapped onto Burkhardt's quartic hypersurface by an explicit mapping (cf. p.337 of [Bu]), but Burkhardt's equations contain a typographical error which was pointed out by Coble (cf. footnote on p.358 of [Co1]).

From our point of view, Burkhardt's mapping can be described in the following way. As in §21, let W^+ and W^- denote respectively the even functions and the odd functions on $K(3,3)$. Then W^+ has dimension 5 and W^- has dimension 4. Furthermore we have

$$W^+ \oplus W^- = V(3,3).$$

We have two Weil representations of $Sp(\mathbf{F}_3^2)$ on $V(3,3)$ corresponding to the two choices, ϵ and ϵ^2, of a primitive cube root of unity. Let us write, for $i = 1, 2$, W_i^+ and W_i^- to denote the invariant subspaces W^+ and W^- for $Sp(\mathbf{F}_3^2)$ in the Weil representation belonging to ϵ^i. We define a homogeneous mapping $\mathcal{B}$ of W_1^- into W_1^+ as the composition

(22.10) $$W_1^- \xrightarrow{i} S^2(W_1^-) \xrightarrow{\mathcal{T}} \bigwedge^2(W_1^+) \xrightarrow{\Pi} \bigwedge^4(W_1^+) \xrightarrow{\tilde{\mathcal{B}}} W_1^+$$

where

(22.11) $$i(w) = w \vee w$$

is the 2-uple embedding, where $\mathcal{T}$ is defined by

(22.12) $$\mathcal{T}f(x,y) = f\left(\frac{x+y}{2}, \frac{x-y}{2}\right)$$

---

[10]   The Hessian of a hypersurface U is the locus of points whose polar quadrics with respect to U are cones. The locus of the singular points of these cones is the Steinerian.

intertwines $Sym^2(W_1^-)$ and $\bigwedge^2(W_1^+)$, where π is defined by

(22.13) $\pi(z) = z \wedge z$

for all z in $\bigwedge^2(W_1^+)$ and where the last mapping in 22.10 is any isomorphism between the two representations. The composed mapping $\mathcal{B}$ is evidently equivariant and has quartic entrics. Hence the preimage under $\mathcal{B}$ of Burkhardt's quartic hypersurface ought to be an invariant hypersurface of degree 16 for $Sp(\mathbf{F}_3^2)$ in $\mathbf{P}(W_1^-)$. But according to Maschke [Mas], there is no invariant of degree 16 for $Sp(\mathbf{F}_3^2)$ in $\mathbf{P}(W_1^-)$. Therefore $\mathcal{B}$ maps $\mathbf{P}(W_1^-)$ onto the Burkhardt quartic. In particular, as Burkhardt already observed, the quartic is uniformized by Siegel modular forms of genus 2. It appears that Coble's work [Co1] shows that the quartic hypersurface is the moduli space for abelian surfaces with level (3,3) structure and trivial Arf invariant. In 1983 Van der Geer [VdG1], who was apparently unaware of Coble's work, showed that Burkhardt's quartic hypersurface is birational to the Siegel modular variety of genus 2 and level 3. His proof is based on his earlier work on certain Hilbert modular surfaces. Our own introduction to Burkhardt's work came from H.F.Baker's wonderful book [Ba] which is devoted to a minute study of the geometry of Burkhardt's quartic.

In each of the above examples, the geometry of the moduli space or abelian variety is closely related to the geometry of some invariant hypersurface for the relevant group. It is not surprising that the group and its invariants should appear because the group does after all act on the moduli space or abelian variety as it is embedded in projective space of the appropriate dimension. We can suppose that the occurrence of such examples was an important part of Felix Klein's motivation for the study of invariants and perhaps for his Erlangen Program as well. However there is still a mystery which is not dispelled by appealing to a general philosophy about the equivalence of the notions of geometry and group action. The geometry which is relevant to our study of the moduli spaces or abelian varieties is the geometry imposed on projective space by the action of the relevant group, being a Heisenberg group or a finite symplectic group or an extension of one by the other. On the other hand, what we observe is that the moduli space is obtained by applying the usual notions of projective geometry to certain invariant hypersurfaces of low degree for the group. These notions include secant loci, singular loci and Hessians in the examples above. However, projective geometry is the geometry imposed on projective space by the group of all collineations, not the geometry imposed by the smaller symmetry groups we are considering. What *is* mysterious, then, is why, starting with one readily available invariant, should the constructions of projective geometry suffice to describe loci and relations belonging to the more refined geometry imposed by the smaller group?

One could dismiss it as fortunate coincidence, but still one has the right to ask when to expect to be favored by coincidence in this way. These issues are discussed in the first author's article [A6], along with pertinent conjectures, and were motivated to some extent by a geometric relationship between two Shimura varieties belonging to different forms of the same arithmetic group discussed in [A2], [A16]. As a concrete question of this type, one can consider the problem of generalizing to higher levels Klein's theorem that the modular curve of level 11 is the singular locus of the Hessian of the cubic threefold 22.4. We have shown that for any prime number p congruent to 3 modulo 8, there is a unique cubic invariant for an irreducible representation of $PSL_2(\mathbf{F}_p)$ acting on $\mathbf{C}^n$ with $2n + 1 = p$. Subsequently the first author wrote the invariant down explicitly and determined its automorphism group [A10]. Recently the existence and explicit form of the cubic invariant were generalized still further in [A5], [A17] (which also correct many regrettable typographical errors in [A10]) to the case of finite fields of odd characteristic and the automorphism groups of the resulting 3 tensors were determined in the general setting. For definiteness, let $p \geq 11$ be a prime congruent to 3 modulo 8. The problem is this: **give an explicit construction of the modular curve of level p from the cubic invariant using projective geometry or prove that no such construction exists.**

Another aspect of the question of the adequacy of constructions of projective geometry to describe relations belonging to a more refined geometry is the problem of generalizing special theorems about invariant loci to theorems about a class of loci having apparently nothing to do with the original moduli problems. For example, the decomposition of the Jacobian variety of the modular curve $X(11)$ into invariant abelian varieties of dimensions 5, 10 and 11 generalizes to a property of cubic threefolds,[11] as explained in [A15], [A16]. Persistence in comparing the geometries in this way enables one to gain unexpected insights into both geometries.

In [A6], the first author considered the question of defining what one means by a construction and stated conjectures some of which imply, among other things, that such a construction always exists. Subsequently, David Vogan read the preprint of [A6] and pointed out how to prove those conjectures that pertained to geometric constructions. With his kind permission, the proofs were incorporated into the paper [A6].

The approach described here involves a careful study of the refined geometry imposed by finite groups of collineations on projective space. Evidently one tool for studying examples is an understanding of the rings of invariants of these finite groups. The problems of computing such rings of invariants and some conjectures regarding their structures as bicycles are also considered in [A6]. In the case of the simple group of order 660 acting on $\mathbf{P}^4$, the first author has computed the full ring of invariants [A8], [A9] and as a simple application has obtained in [A11] equations defining (set theoretically) one of Klein's other embeddings of the modular curve of level 11 in $\mathbf{P}^4$, namely its equivariant projection from its canonical embedding in $\mathbf{P}^{25}$. The paper [A11] is included as Appendix III to this book.

In the paper [A13], the first author showed that in characteristic 3, the Mathieu group M_{11} acts on the modular curve $X(11)$. This follows from Klein's theorem and from the remarkable fact that although the cubic form

$$v^2 w + w^2 x + x^2 y + y^2 z + z^2 v$$

is not invariant under M_{11}, it *is* invariant under M_{11} modulo cubes of linear forms. It is hard to imagine how one could have proved this using the usual methods of studying modular curves. Thus, our notion of "invariant" should be widened so

---

[11] The generalization is not what one would expect at first glance.

as to include this example. Applications of this result will be found in the article
[A15]. One other incidental result, noticed in [A13] and proved in [A15], is that
Klein's A-curve of level 11, which is an embedding of $X(11)$ in $\mathbf{P}^5(\mathbf{C})$ as a curve
of degree 25, is the   singular locus of the unique quartic invariant
for $SL_2(\mathbf{F}_{11})$. Furthermore, that quartic invariant arises in the following way. The
matrix of second partials of the above cubic form can be viewed as a mapping from
V^{-*} to $Sym^2(V^-)$. Identifying $Sym^2(V^-)$ equivariantly with $\bigwedge^2(V^+)$, we obtain
the skew symmetric 6×6 matrix

$$(22.14) \qquad \begin{pmatrix} 0 & v & w & x & y & z \\ -v & 0 & 0 & z & -x & 0 \\ -w & 0 & 0 & 0 & v & -y \\ -x & -z & 0 & 0 & 0 & w \\ -y & x & -v & 0 & 0 & 0 \\ -z & 0 & y & -w & 0 & 0 \end{pmatrix}$$

whose entries are linear forms in v, w, x, y, z. One recovers the original cubic as the
Pfaffian of this skew symmetric matrix, up to a scalar. Each point of the cubic
hypersurface therefore gives rise to a projective line in $\mathbf{P}^5(\mathbf{C})$ and the union of
these lines is the unique quartic invariant of $SL_2(\mathbf{F}_{11})$. There is, of course, also an
invariant rank 2 vector bundle on the cubic threefold, namely the bundle of null
spaces of the above matrix as $[v, w, x, y, z]$ varies over the cubic threefold. Since
the cubic contains (App.II, Lemma 35.1. cf. [K-F]) Klein's degree 50 embedding of
$X(11)$ in $\mathbf{P}^4(\mathbf{C})$, we therefore obtain by restriction yet another equivariant rank 2
vector bundle on the modular curve $X(11)$.

This invariant rank 2 vector bundle on the cubic is an example of a general
phenomenon. Indeed, in [A15], it is shown that the generic cubic threefold admits
a 5 dimensional family of rank 2 vector bundles arising from representations of the
cubic as the Pfaffian of a 6×6 skew symmetric matrix whose entries are linear
forms in 5 variables. Given such a representation, one can consider the family of
null spaces of the matrix as one varies over the cubic threefold. If the rank of the
matrix is never less than 4, the family of null spaces will be a rank 2 vector bundle
on the cubic threefold. Each null space determines a projective line in $\mathbf{P}^5$ and these
projective lines sweep out a quartic hypersurface in $\mathbf{P}^5$.

The singular locus of the quartic consists of a smooth curve of degree 25 and
genus 26, which in the case of the matrix (22.14) is the modular curve $X(11)$, more
precisely Klein's A-curve of level 11. In general, this curve has a canonical mapping
into the Fano surface of lines in the cubic threefold. Its image is the family of
jumping lines of the rank 2 vector bundle. Furthermore, the intermediate jacobian
variety of the cubic threefold is a factor of the jacobian variety of the singular curve
of the quartic. Thus one obtains a family of curves of genus 26 naturally associated
to cubic threefolds, there being ∞^5 curves for each cubic, and for these curves the
jacobian varieties have the intermediate jacobian of the cubic as a factor.

One has an entirely different family of curves of genus 26 associated to cubic
threefolds by considering the nodal curve of the Hessian of the cubic. In general
their jacobians decompose as the sum of abelian varieties of dimensions 10 and 16.
It appears that the jacobian varieties do not decompose further in general and in

[A15] it is shown that any such decomposition cannot be rational over the moduli space of cubic threefolds. The modular curve $X(11)$, remarkably enough, arises from both families and that in part seems to explain the decomposition of its Jacobian variety into $PSL_2(\mathbf{F}_{11})$-invariant subvarieties of dimensions 5,10 and 11, the one of dimension 5 being isomorphic to the intermediate Jacobian of the cubic 22.4

Remark (22.15):   The first author has just been informed that Sorin Popescu and Mark Gross [Po-Gr1], [Po-Gr2] have recently proved that the moduli space of abelian surfaces with level $(1,11)$ structure is birationally equivalent to Klein's cubic. It will be very interesting to reexamine all that has been learned about Klein's cubic in the light of this strikingly beautiful new discovery.

§23   Arithmetic of the Modular Curve as Deduced From Its Equations

In this section, we will show how to compute the degree of the modular curve $\mathcal{C}$ of level $\delta = (p)$ as given by equations (19.13) and give some applications of the methods involved. Here, p is a prime number ≥ 7. Among the applications is a relation between the rational points of the modular curve and rational points on Fermat curves. Such relations are known (cf. [Ku], [Vé]) but the emphasis here is on the elementary nature of our approach, which is modeled on Hurwitz's proof [Hu2] for the case $p = 7$. Unlike the proof of [Vé2], our argument is based on the explicit equations 19.13 and not on the knowledge that these equations are related to moduli of elliptic curves. In [K4], Klein computed the degree using an explicit uniformization by theta functions. The equations define the modular curve as a scheme over $Spec(\mathbf{Z})$.

Let R be a discrete valuation ring with maximal ideal P and with residue field K. Let h be an R-valued point of $\mathcal{C}$ in $\mathbf{P}^m(R)$ where

$$m = (p - 3)/2.$$

We may suppose that

$$h : K(\delta) \to R$$

is an odd function which does not map $K(\delta)$ entirely into P. Suppose however that for some $x \neq 0$ in $K(\delta)$, $h(x)$ lies entirely in P. Then reducing h modulo P we obtain a K-valued point $\widetilde{h}$ of $\mathcal{C}$ vanishing at x. It follows then from Proposition 20.8 that $\widetilde{h}$ is proportional to

$$\phi_a - \phi_{-a}$$

for some a in $K(\delta)$, where for u in $K(\delta)$, the function ϕ_u, as in 21.52, is equal to 1 at u and equal to 0 elsewhere. Similarly, for u in $K(p)$, we denote by

$$\delta_u : K(p) \to R$$

the function from $K(p)$ to R whose value is 1 at u and 0 elsewhere. Then the image in k of δ_u under the canonical mapping of R onto its residue class field k is ϕ_u. In the following theorem let us suppose that $a = 1$ denote by ord the valuation of R with values in $\mathbf{Z}$. The case of general a will be handled in Corollary 23.27.

Theorem (23.1):   *Let R be a discrete valuation ring with maximal ideal P, residue field k and valuation ord. Let h be a function from $K(p)$ to R which reduces modulo P to*

$$\phi_1 - \phi_{-1}.$$

Suppose that h satisfies the equations 19.13. Then we have

(23.2) $$\operatorname{ord} h(2t-1) = \binom{t}{2} \operatorname{ord} h(3)$$

for $1 \le t \le (p-1)/2$. Here we define $\binom{i}{2}$ to be 0.

Proof: For $1 \le t \le (p-1)/2$, let $\mathcal{H}(t)$ denote the statement

(23.2) $$A(t) \text{ and } B(t)$$

where the statements $A(t)$ and $B(t)$ are given by

(23.3)
$$A(t) \;=\; \left[(1 \le i \le t) \Rightarrow \left(\nu_{2i-1} = \binom{i}{2} \nu_3 \right) \right]$$

$$B(t) \;=\; \left[(t+1 \le i \le \tfrac{1}{2}(p-1)) \Rightarrow \left(\nu_{2i-1} > \binom{t}{2} \nu_3 \right) \right]$$

and where

$$\nu_i = \operatorname{ord} h(i)$$

for all i in $K(p)$. The theorem we want to prove is $A((p-1)/2)$. It is therefore sufficient to prove $\mathcal{H}(t)$ for $1 \le t \le (p-1)/2$ by induction on t. The case $\mathcal{H}(1)$ follows at once from the hypothesis that h is congruent to $\phi_1 - \phi_{-1}$ modulo P. We will require some lemmas.

Lemma (23.4):   *Let y be an element of $K(p)$. Then of the three numbers*

(23.5) $$\nu_{2y-1}, \; 3\nu_{y-1} + \nu_{y+1}, \; 3\nu_y + \nu_{y-2}$$

the two smaller ones are equal.

Proof: This follows at once by applying ord to the relation

(23.6) $\Phi_{-1,y,0,y-1} = h(y-1)^3 \cdot h(y+1) - h(1)^3 \cdot h(2y-1) - h(y)^3 \cdot h(y-2) = 0.$

We can now prove that $\mathcal{H}(2)$ is true. Actually, $A(2)$ is a trivial consequence of $A(1)$, so we just have to prove $B(2)$. Let y be an integer such that $3 \le y \le (p-1)/2$ and such that ν_{2y-1} is minimal. If $\nu_{2y-1} > \nu_3$ then $B(2)$ is true. If $\nu_{2y-1} \le \nu_3$ we have by Lemma 23.4 that

(23.7) $$\nu_{2y-1} \ge 3\nu_z + \nu_w$$

where the pair (z, w) is either $(y - 1, y + 1)$ or $(y, y - 2)$. We have

$$\nu_z < \nu_{2y-1},$$

which implies, after writing z in the form $\pm(2u - 1)$ with $1 \le u \le (p - 1)/2$, that $u = 1$ and $z = \pm 1$. But then $y = -2, \pm 1, 0$, which contradicts $3 \le y \le (p - 1)/2$ since $p \ge 7$. This proves $\mathcal{H}(2)$. The proof of $\mathcal{H}(t)$ by induction is similar in spirit but requires some additional lemmas. The following lemma is quite easy and is left to the reader.

Lemma (23.8):   *Let ∇ denote the set of all triangular numbers. Explicitly,*

$$(23.9) \qquad \nabla = \left\{ \binom{s}{2} \;\middle|\; s \ge 1 \right\}.$$

Let ∇' denote the set of all pairs $\binom{r}{2}$, $\binom{s}{2}$ of triangular numbers such that $|r - s| = 1$. Then the function

$$(23.10) \qquad f(x, y) = 3x + y$$

maps the set ∇' bijectively onto the set ∇. Furthermore, we have

$$(23.11) \qquad f\left(\binom{r}{2}, \binom{s}{2} \right) = \begin{cases} \dbinom{2r}{2} & \text{if } r < s \\[2ex] \dbinom{2r + 1}{2} & \text{if } r > s. \end{cases}$$

Lemma (23.12):   *Let $1 < t < (p - 1)/2$ and assume that $\mathcal{H}(t)$ holds. Then we have*

$$(23.13) \qquad \begin{aligned} \nu_{2t+1} \;&=\; \begin{cases} 3\nu_t + \nu_{t+2} & \text{if } t \text{ is odd} \\[1ex] 3\nu_{t+1} + \nu_{t-1} & \text{if } t \text{ is even} \end{cases} \\[2ex] &<\; \begin{cases} 3\nu_{t+1} + \nu_{t-1} & \text{if } t \text{ is odd} \\[1ex] 3\nu_t + \nu_{t+2} & \text{if } t \text{ is even}. \end{cases} \end{aligned}$$

In either case, we have

$$(23.14) \qquad \nu_{2t+1} = \binom{t + 1}{2}.$$

Proof: We have two cases according to whether t is even or odd. Here, the terms "even" and "odd" refer to the parity of a representative of t which is a nonnegative integer less than p. We will prove 23.13 for t odd. So suppose that t is odd and write $t = 2k - 1$. Then since $t > 1$ we have $k + 1 \le t$ and $A(t)$ implies that

$$(23.15) \qquad \nu_t = \nu_{2k-1} = \binom{k}{2} \nu_3$$

and

(23.16)
$$\nu_{t+2} = \nu_{2k+1} = \binom{k+1}{2} \nu_3$$

whence

$$3\nu_t + \nu_{t+2} = \binom{t+1}{2} \nu_3.$$

On the other hand, we have

$$\nu_{t+1} = \nu_{p-t-1} = \nu_{2m-1}$$

and

$$\nu_{t-1} = \nu_{2m+1}$$

where

$$m = -k + (p+1)/2.$$

Since $t > 1$, we have

$$m + 1 \le (p-1)/2.$$

If $m \ge t$ then $\mathcal{H}(t)$ implies that

$$\nu_{t+1} = \nu_{2m-1} \ge \binom{t}{2} \nu_3$$

and

$$\nu_{t-1} = \nu_{2m+1} > \binom{t}{2} \nu_3$$

In that case, we conclude that

(23.17)
$$3\nu_t + \nu_{t+2} = \binom{t+1}{2} \nu_3 < 4 \binom{t}{2} \nu_3 < 3\nu_{t+1} + \nu_{t-1}.$$

If instead $m < t$, we have $m + 1 \le t$ so by $A(t)$ and Lemma 23.8 we have

$$3\nu_{t+1} + \nu_{t-1} = \binom{p-t}{2} \nu_3.$$

Since $t < (p-1)/2$, we have

(23.18)
$$3\nu_t + \nu_{t+2} = \binom{t+1}{2} \nu_3 < 4 \binom{p-t}{2} \nu_3 < 3\nu_{t+1} + \nu_{t-1}.$$

Therefore in either case we have

(23.19)
$$\binom{t+1}{2} = 3\nu_t + \nu_{t+2} < 3\nu_{t+1} + \nu_{t-1},$$

and our assertion for t odd follows from Lemma 23.4. The proof for t even is quite similar and the details are left to the reader.

The reader will note that for $t < (p-1)/2$ the implication

$$\mathcal{H}(t) \Rightarrow A(t+1)$$

follows from Lemma 23.13. It remains to prove that

$$\mathcal{H}(t) \Rightarrow B(t+1)$$

for such t. This is accomplished by the following Lemma.

Lemma (23.20):   *Let $1 < t < (p-1)/2$ and assume $\mathcal{H}(t)$. Then $B(t+1)$ is true.*

Proof: Let

(23.21) $$M = \{j \mid t+1 \le j \le \tfrac{1}{2}(n-1) \},$$

let

(23.22) $$\mu = \inf \{\nu_{2y-1} \mid y \in M\}$$

and let

(23.23) $$N = \{y \in M \mid \nu_{2y-1} = \mu\}.$$

By Lemma 23.13, the statement $A(t+1)$ holds. So

$$\mu \le \nu_{2t+1} = \binom{t+1}{2} \nu_3.$$

Let y be an element of N. Let α denote the value of $3\nu_z + \nu_w$ for $(z, w) = (y-1, y+1)$ and let β denote its value for $(z, w) = (y, y-2)$. By Lemma 23.4 we have

$$\mu \ge \alpha$$

or

$$\mu \ge \beta.$$

If either of these inequalities is strict then we will have

$$\mu > \alpha = \beta.$$

But the inequality

$$\mu \ge 3\nu_z + \nu_w$$

implies

$$\mu > \nu_z$$

and by $\mathcal{H}(1)$ and the fact that $n \ge 7$ and y belongs to N, we have

$$\nu_z > 0$$

and

$$\mu > \nu_w$$

as well. By definition of μ and by $\mathcal{H}(t)$ we then have that the pair

$$(\nu_z/\nu_3, \nu_w/\nu_3)$$

lies in ∇' where ∇' is as defined in Lemma 23.8. Therefore, if

$$\mu > \alpha = \beta,$$

we have by Lemma 23.8 that

$$\nu_y - 1 = \nu_y$$

and

$$\nu_{y+1} = \nu_{y-2}.$$

Since all of these are less than μ, $\mathcal{H}(t)$ implies that

$$y - 1 \equiv \pm y \pmod{p},$$

which contradicts the fact that y belongs to M. So the relation $\mu > \alpha = \beta$ cannot hold. Therefore by Lemma 23.4 we have either $\mu = \alpha$ or $\mu = \beta$. It follows that

$$\mu = 3\nu_z + \nu_w,$$

where $(z, w) = (y - 1, y + 1)$ or $(z, w) = (y, y - 2)$, and reasoning as above we have

$$\mu > \nu_z$$

and

$$\mu > \nu_w.$$

By $\mathcal{H}(t)$ and the definition of μ we can write

(23.24)
$$z \equiv \pm(2u - 1) \pmod{p}$$
$$w \equiv \pm(2v - 1) \pmod{p}$$

with $1 \leq u, v \leq t$. We then have

(23.25)
$$\nu_z = \binom{u}{2}\nu_3$$
$$\nu_w = \binom{v}{2}\nu_3.$$

By Lemma 23.8 we have

$$3\nu_z + \nu_w = \binom{x}{2}\nu_3,$$

where

$$x = \begin{cases} 2u & \text{if } u < v \\ 2u - 1 & \text{if } u > v. \end{cases}$$

In either case, we have

$$\mu = \binom{x}{2}\nu_3,$$

but since we have already observed that

$$\mu \le \binom{t+1}{2}\nu_3,$$

we conclude from $B(t)$ that $x = t + 1$ and therefore that

$$\mu = \binom{t+1}{2}\nu_3$$

and

$$t = x - 1 \ge 2u - 2.$$

I claim that z must be odd. For if z is even then

$$z = p - (2u - 1)$$

and, since

$$y \ge z$$

and

$$t \ge 2u - 2,$$

we would then have

$$y \ge p - (2u - 1) \ge p - t - 1,$$

contradicting the fact that y belongs to M. It follows that if y is even then

$$z = y - 1$$
$$u < v$$
$$t = 2u - 1$$

and therefore that

$$y = t + 1.$$

On the other hand, if y is odd then

$$z = y, \ u > v, \ t = 2u$$

and again $y = t + 1$. This proves that the only element of N is $t + 1$ and $B(t + 1)$ is completely proved.

Corollary (23.26):   *We adopt the assumptions and conventions of Theorem 23.1 except that we now suppose that*

$$h \equiv \phi_a - \phi_{-a} \pmod{P},$$

where a is any nonzero element of $K(p)$. Then we have

(23.27) $$\operatorname{ord} h\big((2t+1)a\big) = \binom{t}{2} \operatorname{ord} h(3a).$$

Proof: Apply the automorphism

$$\alpha : a \mapsto ax$$

to the additive group $K(p)$. Then α acts by composition on $V(p)$ and one sees very easily that α preserves the system of equations 19.13 and moves $\phi_{\pm 1}$ to $\phi_{\pm a}$. It therefore transforms Equation 23.2 into Equation 23.28 and we are done.

Corollary (23.28):   *Let $\delta = (p)$. The curve part C of the locus $\mathcal{L}$ defined by 19.13 has degree*

(23.29) $$\frac{(p-3)\cdot(p-1)\cdot(p+1)}{48} = \binom{\frac{p+1}{2}}{3}$$

Proof: As in §20, for every u in $K(p)$ denote by E_u the linear functional on V^- given by

$$E_u(h) = h(u)$$

for every h in V^-. Then we have

$$E_{-u} = -E_u$$

and

$$E_0 = 0$$

For each $a \neq 0$ in $K(\delta)$, let κ_a be the point of C represented by $\phi_a - \phi_{-a}$. Consider the local ring R_a of C at a. By Corollary 20.29, the point κ_a is a simple point of C and therefore R_a is a discrete valuation ring. We denote its valuation by ord_a and its maximal ideal by P_a. For every u in $K(p)$, the quotient E_u/E_a determines an element of the local ring R_a. If we associate to each u in $K(p)$ the element of R_a obtained in this way, we obtain a function h_a from $K(p)$ into R_a. We will write

(23.30) $$h_a(u) = \frac{E_u}{E_a}.$$

Then h_a is an odd function on $K(p)$ with values in the discrete valuation ring R_a. The canonical mapping of R_a onto its residue field is simply evaluation at κ_a. We then have

(23.31) $$h_a(u)(\kappa_a) = \phi_a(u) - \phi_{-a}(u)$$

which shows that
$$h_a \equiv \phi_a - \phi_{-a} \pmod{P_a}.$$
Therefore, by taking
$$R = R_a$$
$$P = P_a$$
$$\text{ord} = \text{ord}_a$$
in Corollary 23.27, we have

(23.32)
$$\text{ord}_a \, h_a \left((2t - 1) \, a \right) = \binom{t}{2} \, \text{ord}_a \, h_a(3a)$$

for $1 \leq t \leq (p-1)/2$. As we have already observed in Corollary 20.29, the point κ_a is a simple point of the curve $\mathcal{C}$. Since the linear mappings E_u with $u \neq \pm a$ span all of the linear functionals on V^- vanishing at κ_a, it follows that for some u in $K(p)$, the rational function h_a on $\mathcal{C}$ must vanish to order 1 at κ_a. By equation 23.33,

$$\text{ord}_a \, h_a(3a) = \inf_{u \neq \pm a} \text{ord}_a \, h_a(u),$$

so we must have
$$\text{ord}_a \, h_a(3a) = 1.$$
Therefore, for all $t \neq 0$ in $K(p)$, we have

(23.33)
$$\text{ord}_a \, h_a \left((2t - 1) \, a \right) = \binom{t}{2}$$

for $1 \leq t \leq (p-1)/2$. Consider the hyperplane H in $\mathbf{P}(V^-)$ given by $E_1 = 0$. By Proposition 20.8, H meets $\mathcal{C}$ only at the points κ_{2u-1} with $1 \leq u \leq (p-1)/2$. The order to which H meets $\mathcal{C}$ at κ_a is equal to the order to which $E_1/E_q = h_a(1)$ vanishes at a or, what is the same, $\text{ord}_a \, h_a(1)$. Replacing a by $-a$ if necessary, which will not change κ_a, we can find an integer t_a such that $1 \leq t_a \leq (p-1)/2$ and such that

(23.34)
$$(2t_a - 1) \, a \equiv 1 \pmod{p}.$$

We then have that H meets $\mathcal{C}$ to order $\binom{t_a}{2}$ at κ_a. As we run over all of the points κ_a the integers t_a will run over all of the values between 1 and $(p-1)/2$. Therefore the degree of $\mathcal{C}$ is simply

(23.35)
$$\sum_{t=1}^{\frac{1}{2}(p-1)} \binom{t}{2} = \frac{(p-3) \cdot (p-1) \cdot (p+1)}{48},$$

and we are done.

The points κ_a are K-rational for any field K. They are called the **trivial** K-rational points of $\mathcal{C}$. Similarly if we take a "twisted" Fermat curve of the form

$$\alpha \cdot x^p + \beta \cdot y^p + \gamma \cdot z^p = 0$$

with α, β, γ in K, a K-rational point of the Fermat curve with $xyz = 0$ is called a **trivial** K-rational point of the Fermat curve.

Corollary (23.36):   *Let $p \geq 7$ be a prime number and let K be a number field whose class number is not divisible by p. Suppose that C has a nontrivial K-rational point. Then we can find units α, β and γ of K such that the "twisted" Fermat curve*

$$(23.37) \qquad\qquad \alpha \cdot x^p + \beta \cdot y^p + \gamma \cdot z^p = 0$$

has a nontrivial K-rational point.

Proof: Let $[h]$ be a nontrivial K-rational point of C, where $[h]$ is represented by a function $h : K(p) \to K$ which we may assume has its values in the ring $\mathcal{O}_K$ of integers of K. If q is any prime ideal of K, denote by R_q the q-adic completion of $\mathcal{O}_K$. We can then view $[h]$ as an R_q-valued point of C. If we then divide h by a suitable power of the prime element of R_q, we can assume that h has its values in R_q and that one of its values is a unit of R_q. For definiteness, let a be a nonzero element of $K(p)$ and suppose that $h(a)$ is a unit. If we reduce modulo q, we obtain a point of C rational over the residue class field $\mathcal{O}_K/q$. By Corollary 23.27, if some value of h lies in qR_q then the only nonzero values of h modulo q occur at $\pm a$. In particular, either $h(b)$ is a unit of R_q for all nonzero b in $K(p)$ or else $h(3a)$ is a nonunit. Therefore by Corollary 23.27 we have in either case that

$$(23.38) \qquad\qquad \operatorname{ord}_q h((2t-1)\,a) = \binom{t}{2} \operatorname{ord}_q h(3a)$$

whenever t is a positive integer less than $p/2$. Now in the equations 19.13, let us take

$$w = -1,\; x = 0,\; y = b$$

and

$$z = b - 1,$$

where b is an element of $K(p)$ other than 1, -1, 2 or $(p+1)/2$. Since $p \geq 7$, we can find such a b. We then obtain the equation

$$(23.39) \qquad 0 = h(1)^3 h(2b-1) + h(b-1)^3 h(b+1) - h(b)^3 h(b-2).$$

If c is a positive integer less than $p/2$ then

$$2c - 1 \equiv \pm(2t-1)a \pmod{q}$$

with $1 \leq t \leq p/2$. Since h is an odd function, Theorem 23.1 implies that

$$(23.40) \qquad\qquad \operatorname{ord}_q h(2c-1) = \binom{t}{2} \operatorname{ord}_q h(3a).$$

Modulo p we have that

$$(23.41) \qquad\qquad \binom{t}{2} \equiv \frac{1}{8}\left[-1 + \frac{(2c-1)^2}{a^2} \right].$$

Therefore if b is positive and less than $p/2$, we have modulo p that

(23.42)
$$ord_q\left[h(1)^3h(2b-1)\right] \equiv \left\{\frac{3}{8}\left[-1+\frac{1}{a^2}\right]+\frac{1}{8}\left[-1+\frac{(2b-1)^2}{a^2}\right]\right\} ord_q\, h(3a)$$

(23.43)
$$ord_q\left[h(b-1)^3h(b+1)\right] \equiv \left\{\frac{3}{8}\left[-1+\frac{(b-1)^2}{a^2}\right]+\frac{1}{8}\left[-1+\frac{(b+1)^2}{a^2}\right]\right\} ord_q\, h(3a)$$

and

(23.44) $\quad ord_q\left[h(b)^3h(b-2)\right] \equiv \left\{\frac{3}{8}\left[-1+\frac{b^2}{a^2}\right]+\frac{1}{8}\left[-1+\frac{(b-2)^2}{a^2}\right]\right\} ord_q\, h(3a).$

The reader can easily verify that the right hand sides of 23.43-45 are all congruent modulo p to

(23.45)
$$\frac{1}{8}\, ord_q\, h(3a)\left[-1+\frac{(2b-1)^2}{a^2}\right]$$

and therefore to each other. Furthermore, if we replace h by any constant multiple of h, we will not affect this congruence. Since the prime ideal q of $\mathcal{O}_K$ was arbitrary, we have therefore shown that for any K-rational point $[h]$ of C, the ord_q of the numbers

(23.46) $\qquad h(1)^3h(2b-1) \qquad h(b-1)^3h(b+1) \qquad h(b)^3h(b-2)$

are all congruent modulo p. Denote these three numbers by A, B and C respectively. Then the fractional ideals generated by A/C and B/C must be of the form I_1^p and I_2^p respectively, where I_1 and I_2 are fractional ideals of K. Since we assumed at the outset that p did not divide the class number of K, we may conclude that both of the ideals I_1 and I_2 are principal. We can therefore find elements x, y, z of $\mathcal{O}_K$ such that x/z is a generator of I_1 and y/z is a generator of I_2. Then we can find units α and β of $\mathcal{O}_K$ such that

(23.47)
$$\frac{A}{C} = \alpha\left(\frac{x}{z}\right)^p$$
$$\frac{B}{C} = \beta\left(\frac{y}{z}\right)^p.$$

Since $A + B = C$, we conclude that for $\gamma = 1$ we have

(23.48) $\qquad\qquad\qquad \alpha x^p + \beta y^p + \gamma z^p = 0.$

If the K-rational point $[x, y, z]$ of the curve 23.49 is trivial then it is easy to verify that the point $[h]$ is a trivial K-rational point of C. Details are left to the reader.

Remark (23.49): In case $K = \mathbf{Q}$, the only units are ± 1, which can be absorbed in the variables. So the nontrivial rational points of C determine nontrivial rational solutions of the Fermat equation

$$x^p + y^p + z^p = 0.$$

We conclude that if Fermat's last theorem is true for the prime $p \geq 7$, then $\mathcal{C}$ has no nontrivial rational points. Our argument is modelled on Hurwitz's proof [Hu2] for the special case $p = 7$. Our general argument could certainly have been discovered by Hurwitz since all of the techniques we used are elementary and depend only on the equations 19.13 of Klein [K1-4], which were certainly known to Hurwitz. For the case $K = \mathbf{Q}$, Vélu obtained the same result with p replaced by a prime power using results of Deligne-Rapaport [De-Ra] and otherwise proceeding along similar lines. It is very likely that his methods will improve our Corollary 23.37. It is natural to ask whether there might be an analogue of Kummer's theorem for the modular curve for regular primes. The first step seems to be to investigate the solutions in the p-th cyclotomic field K_p of the equation 23.38 where α, β and γ are units of K_n. If we adjoin p-th roots of α/γ and β/γ we obtain an extension L of K_n in which the equation becomes the ordinary Fermat equation $x^p + y^p + z^p = 0$. Furthermore, by a result of Iwasawa [12], if p does not divide the class number of K_n then it will not divide the class number of L either. Finally, we should point out that Mazur [Ma] has determined all of the rational points on all of the modular curves without relying on Fermat's last theorem. It would be very interesting to study his methods from the elementary point of view adopted here.

§24   Invariant Vector Bundles and Modular Forms of Fractional Weight

We denote by $X(p)$ the modular curve of prime level p over the field of complex numbers. We denote by G the group $PSL_2(\mathbf{F}_p)$ acting on $X(p)$.

Theorem (24.1):   *The group $\mathcal{L}$ of G-invariant line bundles on $X(p)$ is an infinite cyclic group generated by a G-invariant line bundle of degree $\frac{p^2-1}{24}$. The subgroup $\mathcal{L}'$ of $\mathcal{L}$ consisting of line bundles associated to G-invariant divisors on $X(p)$ has index 2 in $\mathcal{L}$. In particular, an invariant line bundle is determined up to isomorphism by its degree.*

Proof: Denote by K the function field of $X(p)$ and by $K^\times$ the multiplicative group of K. Let $A = K^\times/\mathbf{C}^\times$, let B denote the group of all divisors on $X(p)$ and let C denote the group of all isomorphism classes of line bundles on $X(p)$. Then we have the exact sequence

$$1 \to A \to B \to C \to 1.$$

From the associated long exact cohomology sequence of G, we have

$$1 \to A^G \to B^G \to C^G \to H^1(G, A).$$

---

[12] We are indebted to Paul Monsky for bringing Iwasawa's result to our attention.

From the exact sequence

$$1 \to \mathbf{C}^\times \to K^\times \to A \to 1,$$

we deduce that the sequence

$$H^1(G, K^\times) \to H^1(G, A) \to H^2(G, \mathbf{C}^\times)$$

is exact. However, $H^1(G, K^\times)$ is trivial by Hilbert's Theorem 90 and $H^2(G, \mathbf{C}^\times)$ is the group of Schur multipliers of G, which is known (e.g., see Table 4.1, p.302 of [Go1]) to be cyclic of order 2. Since $\mathcal{L} = C^G$ and $\mathcal{L}' = B^G$, it follows that $\mathcal{L}'$ has index ≤ 2 in $\mathcal{L}$. Now, in the work of Felix Klein [Kl2], p.195, there are equivariant embeddings of the modular curve into $\mathbf{P}^r(\mathbf{C})$ and $\mathbf{P}^{r+1}(\mathbf{C})$, where $r = (p-3)/2$, as curves of degrees $(p-3)(p-1)(p+1)/48$ and $(p-1)^2(p+1)/48$ respectively. These were called respectively the z-curve and the A-curve. The difference of these two degrees is $(p^2 - 1)/24$, which proves that $\mathcal{L}$ contains an element of degree $(p^2 - 1)/24$. On the other hand, let $N = (p^3 - p)/2$ denote the order of G. It is well known that the G-orbits of $X(p)$ have orders N, $N/2$, $N/3$ or N/p and no other orders occur. The greatest common divisor of these numbers is $(p^2 - 1)/12$, which implies that $\mathcal{L}' \neq \mathcal{L}$. Therefore $\mathcal{L}'$ has index 2 in $\mathcal{L}$, the degrees of the elements of $\mathcal{L}'$ are the multiples of $(p^2 - 1)/12$ and the degrees of the elements of $\mathcal{L}$ are the multiples of $(p^2 - 1)/24$. It remains to show that an invariant line bundle is uniquely determined by its degree. For that it is enough to show that an invariant line bundle of degree 0 is trivial. Since $\mathcal{L}'$ has index 2 in $\mathcal{L}$ and since the group of degrees of invariant divisors has index 2 in the group of degrees of invariant line bundles, it follows that any invariant line bundle of degree 0 represents an element of $\mathcal{L}'$. Therefore we will be done if we can show that any invariant divisor δ of degree 0 is the divisor of a function. Denote by (∞) the cuspidal divisor, by (i) the divisor of points fixed by involutions of G and by (ω) the divisor of points fixed by elements of order 3 of G. Then we can write δ in the form

$$\delta = a \cdot (\infty) + b(i) + c(\omega) + \sideset{}{'}\sum d_x G \cdot x,$$

where a, b, c, d_x are integers and where $\sum'$ denotes a sum running over a finite number of points x of $X(p)$ lying in distinct orbits $G \cdot x$ other than (∞), (i) or (ω). Since δ has degree 0, we have

$$\frac{a}{p} + \frac{b}{2} + \frac{c}{3} + \sideset{}{'}\sum d_x = 0.$$

It follows that a must be divisible by p, b by 2 and c by 3. The quotient of $X(p)$ by G is isomorphic to $\mathbf{P}^1(\mathbf{C})$ and the natural mapping of $X(p)$ onto that quotient may therefore be regarded as a rational function f on $X(p)$. It follows that the divisors $p \cdot (\infty)$, $2 \cdot (i)$, $3 \cdot (\omega)$, and $G \cdot x$ are all linearly equivalent, where x is any point of $X(p)$ not lying in (∞), (i) or ω. It follows at once that δ is the divisor of a function and the theorem is proved.

Corollary (24.2):   Let $p \geq 5$ be a prime number. Then the only point of the Jacobian variety of $X(p)$ fixed by G is the identity. In particular, if the integral *representation of G on $H_1(X(p), \mathbf{Z})$ is reduced modulo a prime ℓ, the the result-ing ℓ-modular representation of G does not contain the trivial representation as a submodule.*

Corollary (24.3):   *Let $p \geq 5$ be a prime number. Denote by D the divisor on $X(p)$ given by*

$$(24.3.1) \qquad\qquad D = \begin{cases} (i) - (\omega) - n(\infty) & \text{if } p = 6n+1 \\ n(\infty) - (i) + (\omega) & \text{if } p = 6n-1. \end{cases}$$

Then D is a G-invariant divisor whose associated line bundle generates the group $\mathcal{L}'$.

Proof: We easily compute the degree of D to be $(p^2 - 1)/12$. The corollary now follows from Theorem 24.1.

Corollary (24.4):   *Let $p \geq 5$ be a prime number. Let λ denote the generator of $\mathcal{L}$ of degree $(p^2 - 1)/24$ and let κ denote the element of $\mathcal{L}$ corresponding to the canonical line bundle of $X(p)$. Then $\kappa = \lambda^{2p-12}$. In particular, κ belongs to $\mathcal{L}'$ and is associated to the divisor $(p - 6) \cdot D$, where D is the divisor defined in equation (24.3.1).*

Proof: It is well known that the canonical divisor of $X(p)$ has degree

$$\frac{(p^2 - 1)(p - 6)}{12}.$$

From this, the theorem and Corollary 24.3, the result follows at once.

Corollary (24.5):   *Let $p \geq 5$ and let $r = (p - 3)/2$. Denote by α, ζ the line bundles on $X(p)$ associated to the embedding of $X(p)$ as the A-curve in $\mathbf{P}^{r+1}(\mathbf{C})$ and as the z-curve in $\mathbf{P}^r(\mathbf{C})$ respectively. Then we have*

$$\alpha = \lambda^{r+1}$$
$$\zeta = \lambda^r.$$

In particular, exactly one of the elements α, ζ belongs to $\mathcal{L}'$. That element is α if $p \equiv 1 \pmod 4$ and is ζ if $p \equiv 3 \pmod 4$

Proof: This follows at once from the degrees of the A-curve and the z-curve, which are given by [K-F] (for the z-curve, see Corollary 23.28).

In view of the fact that the canonical line bundle on $X(p)$ has a canonical N-th root λ, where $N = 2p - 12$, it is reasonable[13] to refer to the holomorphic sections of λ^r, for $r \in \mathbf{Z}$, as **modular forms of weight** $r/(p-6)$. The idea of using Klein's embeddings of modular curves to define and study certain modular forms of fractional weight was first suggested in [A2]. Thus, Klein's z-curve gives rise to modular forms of weight $(p - 3)/(2p - 12)$ for $\Gamma(p)$ and Klein's A-curve gives rise

---

[13] It must however be pointed out that this notion of modular forms of fractional weight apparently doesn't lend itself to an adelic treatment based on adelic meta-plectic groups.

to modular forms of weight $(p - 1)/(2p - 12)$. In Appendix 2, we will present the unpublished article [A14], in which an explicit basis for the modular forms of weight 4/5 for $\Gamma(11)$ is given.

It is natural to wish to extend our results on $SL_2(\mathbf{F}_p)$ invariant line bundles on $X(p)$ to $SL_2(\mathbf{F}_p)$ invariant vector bundles on $X(p)$. We have many examples of such vector bundles which arise naturally. For example, the natural inclusion of $X(p)$ into the Grassmannian Gr in $\mathbf{P}(\bigwedge^2(V^+))$ gives rise to an invariant rank 2 vector bundle $\mathcal{E}$ which we will describe in more detail below. However, first we begin with some general results.

Proposition (Ramanan) (24.6):   *Let $\mathcal{G}$ be a finite group which acts on a vector bundle $E \to X$ over a curve X. Denote by $\mathcal{H}$ the subgroup of $\mathcal{G}$ acting trivially on X and assume that $\mathcal{H}$ lies in the center of $\mathcal{G}$. For every character ψ of $\mathcal{H}$, denote by $E(\psi)$ the ψ-subbundle of E, i.e. the maximal subbundle[14] of E such that $h \cdot e = \psi(h)e$ for all e in $E(\psi)$ all $h \in \mathcal{H}$. Assume that for every character ψ of $\mathcal{H}$ for which $E(\psi)$ is nonzero, there exists a $\mathcal{G}$-invariant line bundle M on X such that $M(\psi) = M$. Then if r is the rank of E then there exists a flag*

$$0 \subset E_1 \subset E_2 \subset \ldots \subset E_{r-1} \subset E$$

of $\mathcal{G}$-invariant subbundles of E, where $rank(E_i) = i$.

Proof: It suffices to show that if $r > 1$ then E has a nonzero $\mathcal{G}$-invariant subbundle F other than E. Therefore, let us assume the contrary, i.e. that E is $\mathcal{G}$-simple, and obtain a contradiction. Our assumptions imply that for some character ψ of $\mathcal{H}$, the subbundle $E(\psi)$ is nonzero. Since we assume that E has no nonzero $\mathcal{G}$-invariant subbundle of smaller rank, it follows that $E = E(\psi)$. By hypothesis, there is a $\mathcal{G}$-invariant vector bundle M such that $M = M(\psi)$. Then $\mathcal{H}$ acts trivially on $M^* \otimes E$ and it will suffice to show that $M^* \otimes E$ is not $\mathcal{G}$-simple. Therefore, by replacing E by $M^* \otimes E$ we can assume that the subgroup $\mathcal{H}$ is trivial. Let L be an ample line bundle on X. By replacing L by the tensor product of all of its pullbacks via elements of $\mathcal{G}$, we can assume that L is $\mathcal{G}$-invariant. By replacing L by a suitable power of L, we can assume that at any point x of X, the isotropy group $\mathcal{G}_x$ of x in $\mathcal{G}$ acts trivially on the fibre L_x of L at x. Let n be an integer. Applying the Woods Hole Fixed Point Formula (SGA 5 III.6.1.2, p.131) to the vector bundle $L^n \otimes E$, we find that for all $g \in \mathcal{G}$ such that g acts nontrivially on X, we have

$$(24.7) \quad tr(g \mid H^0(X, L^n \otimes E)) - tr(g \mid H^1(X, L^n \otimes E)) = \sum_{g(x)=x} \frac{tr(g \mid L_x^n \otimes E_x)}{1 - dh_x}.$$

Since g acts trivially on L_x^n, the right hand side is independent of n. If $n >> 0$, the space $H^1(X, L^n \otimes E) = 0$ since L is ample and the left hand side reduces to $tr(g \mid H^0(X, L^n \otimes E))$. Ampleness of L also implies that the dimension of $H^0(X, L^n \otimes E)$ goes to infinity with n. Denote by χ_n the character of $\mathcal{G}$ on $H^0(X, L^n \otimes E)$. Then

---

[14] To see that this subbundle exists, note that there exists a Zariski open subset U of X such that $\{e \in E \mid h \cdot e = \psi(h)e\}$ induces a vector bundle over U. This vector bundle then extends uniquely to a subbundle of E over all of X.

for $m, n >> 0$, the virtual character $\chi_m - \chi_n$ vanishes on all elements of $\mathcal{G}$ which act nontrivially on X, i.e. all elements of $\mathcal{G}$ other than the identity element. It follows that $\chi_m - \chi_n$ is a sum of copies of the regular representation of $\mathcal{G}$. In particular, for large n, the representation of $\mathcal{G}$ on $H^0(X, L^n \otimes E)$ must contain a copy of the trivial representation. Hence there is a nontrivial $\mathcal{G}$-equivariant homomorphism of L^{-n} into E. The image might not be a vector bundle, but in any case the image will generate a $\mathcal{G}$-invariant line subbundle, and we are done.

Corollary (24.8): *Let E be a $SL_2(\mathbf{F}_p)$-invariant vector bundle of rank r on the modular curve $X(p)$. Then there is a flag*

$$0 \subset E_1 \subset E_2 \subset \ldots \subset E_{r-1} \subset E$$

of $SL_2(\mathbf{F}_p)$-invariant subbundles of E, where $rank(E_i) = i$.

Proof: The subgroup of $SL_2(\mathbf{F}_p)$ acting trivially on $X(p)$ is the center, which has order 2. Since the center acts nontrivially on the invariant line bundle λ of Cor.24.4, the hypotheses of Theorem 24.6 are satisfied and we are done.

We do not know how to find an explicit $PSL_2(\mathbf{F}_p)$ invariant flag for the $PSL_2(\mathbf{F}_p)$ invariant rank 2 vector bundle $\mathcal{E}$ on $X(p)$ associated to the inclusion of the modular curve into the Grassmannian Gr by pulling back the tautological bundle on Gr. We can, however, find a B_p-invariant flag, where B_p is defined in Lemma 24.11 below.

Proposition (Ramanan) (24.9): *Denote by $\mathcal{E}$ the vector bundle on $X(p)$ associated to the natural inclusion of $X(p)$ into the Grassmannian Gr. Then we have an exact sequence*

(24.10) $$0 \to \zeta^{-2} \to \mathcal{E} \to \mathbf{1} \to 0.$$

where $\mathbf{1}$ denotes the trivial line bundle on $X(p)$.

Proof: For any point $[f]$ of $X(p)$ in $\mathbf{P}(V^-)$, corresponding to an odd function $f : K(p) \to k$, The function f^2 is an even function, hence determines a point $[f^2]$ of $\mathbf{P}(V^+)$. It follows from Equation 20.9, with $y = 0$, the remarks following it that the point $[f^2]$ lies on the line in $\mathbf{P}(V^+)$ associated to the decomposable element $F \wedge G$ of $\bigwedge^2(V^+)$. We therefore have a canonical embedding of ζ^{-2} into $\mathcal{E}$. Since in Equation 20.9 we can replace F by ϕ^2, it follows that the cokernel of $\zeta^{-2} \to \mathcal{E}$ has a nowhere vanishing global section represented by the Dirac delta f_0 of 0 on $K(p)$, and is therefore the trivial line bundle. This proves the sequence exact.

Thus, $\mathcal{E}$ is an extension of $\mathbf{1}$ by ζ^{-2}. This extension also has certain equivariance properties. Roughly speaking, it is equivariant with respect to the group of upper triangular matrices of $PSL_2(\mathbf{F}_p)$. Now, if p is congruent to 3 modulo 4 then the group $PSL_2(\mathbf{F}_p)$ does not act on V^+; it is $SL_2(\mathbf{F}_p)$ that acts. Furthermore, the vector bundle $\mathcal{E}$ is contained in a trivial V^+ bundle, so for such p, the group $SL_2(\mathbf{F}_p)$ acts on $\mathcal{E}$, not the group $PSL_2(\mathbf{F}_p)$, and the center of $SL_2(\mathbf{F}_p)$ acts by multiplication by -1. However, the center of $SL_2(\mathbf{F}_p)$ is contained in the group of upper triangular matrices in $SL_2(\mathbf{F}_p)$ but that center also acts trivially on the $SL_2(\mathbf{F}_p)$-invariant line bundles ζ^{-2} and $\mathbf{1}$. Therefore, we must take some care in formulating the invariance properties of this exact sequence in order to avoid certain apparent paradoxes.

Lemma (24.11): *Denote by B the upper triangular subgroup of $SL_2(\mathbf{F}_p)$ and by $\overline{B}$ the image of B in $PSL_2(\mathbf{F}_p)$ under the natural homomorphism π of $SL_2(\mathbf{F}_p)$ onto $PSL_2(\mathbf{F}_p)$. If p is congruent to 3 modulo 4, the homomorphism π admits a canonical splitting over $\overline{B}$, which we denote σ. We will write B_p to denote B if p is congruent to 1 modulo 4 and to denote $\overline{B}$ if p is congruent to 3 modulo 4. Similarly, we will write π_p to denote π or σ according to whether p is 1 or 3 modulo 4. Then the exact sequence 24.10 is B_p-equivariant via π_p. In the two cases, $SL_2(\mathbf{F}_p)$ is understood to be acting on V^- via the representation ρ_t^- and on V^+ via the representation ρ_{2t}^+.*

Proof: This is straightforward and left to the reader.

Ramanan also determined the explicit cocycle of this extension.

Proposition (Ramanan) (24.12): *For each nonzero element a of $K(p)$, denote by U_a the complement of the locus $E_a = 0$ in $X(p)$. Then the U_a form a covering of $X(p)$. With respect to this covering, the extension 24.10 is defined by the cocycle*

$$(24.13) \qquad c(U_a, U_b)[f](x) = \frac{f(x)^2 f(a+b) f(a-b)}{f(a)^2 f(b)^2}.$$

This expression is to be interpreted in the following way: the line bundle ζ^{-2} is a subbundle of the trivial V^+ bundle on $X(p)$, so its sections over an open set U can be regarded as V^+-valued functions on U.

Proof: On each U_a, consider the expression

$$-\frac{f(x+a)f(x-a)}{f(a)^2},$$

which we denote by $G_a^f(x)$ when we wish to emphasize that it is an even function from $K(p)$ to k, and which we denote $G_a(f, x)$ or $G_a([f], x)$ when we wish to regard it as a function of $[f] \in U_a$ with values in V^+. It is then easy to see that, for $[f] \in U_a$, the space spanned by G_a^f and f^2 is the image of $[f]$ by the map from $X(p)$ into the Grassmannian Gr. In other words, we claim that

$$f(x+y)f(x-y) = (f^2 \wedge G_a^f)(x,y) = f^2(x)G_a^f(y) - f^2(y)G_a^f(x)$$

for all f in V^- such that $[f] \in U_a$. But this is easily verified by multiplying both sides by $f(a)^2$, transposing the terms on the right hand side to the left and observing that the resulting identity is $-\Phi_{0axy} = 0$. It follows that G_a is a section of $\mathcal{E}$ over U_a and defines a splitting of the exact sequence 24.10 over U_a. In order to obtain the explicit cocycle, we have to take the coboundary $c = \delta c_0$ of the $\mathcal{E}$-valued 0-cochain $c_0 : U_a \mapsto G_a$. We then have

$$c(U_a, U_b) = \delta c_0(U_a, U_b) = c_0(U_b) - c_0(U_a) = G_a - G_b.$$

For $[f] \in U_a \cap U_b$, $c(U_a, U_b)([f])$ belongs to V^+, since $\mathcal{E}$ is contained in the trivial V^+ bundle on $X(p)$, and we have

$$c(U_a, u_b)([f])(x) = \frac{f(b)^2 f(x+a) f(x-a) - f(a)^2 f(x+b) f(x-b)}{f(a)^2 f(b)^2}$$

for $x \in K(p)$. However, using the quartic relation $\Phi_{0xab} = 0$, this expression becomes

$$c(U_a, U_b)([f])(x) = \frac{f(x)^2 f(a+b) f(a-b)}{f(a)^2 f(b)^2},$$

which we may regard as a section of ζ^{-2} over $U_a \cap U_b$. This expression also shows why, f^2 the G_a piece together to give the unique section of the quotient. This gives the required element of $H^1(\zeta^{-2})$.

Lemma (Ramanan) (24.14): *The exact sequence 24.10 does not split if $X(p)$ is not a rational curve.*

Proof: If the exact sequence splits then the bundle $\mathcal{E}$ has a nowhere vanishing section. In that case, since $\mathcal{E}$ is a subbundle of the trivial V^+ bundle on $X(p)$, there exists a vector $v \in V^+$, independent of $[f] \in X(p)$, such that

$$f(x+y) f(x-y) = (f^2 \wedge v)(x, y).$$

Taking $[f]$ to be generic on $X(p)$ and taking $y = 1$, and successively, solving for $f(x)$, we see by induction that f is determined by $f(1)$ and $f(2)$. This contradicts the assumption that $X(p)$ is not a rational curve.

We note that since the full group $SL_2(\mathbf{F}_p)$ does not act monomially, there is no reason to expect the exact sequence 24.10 to be equivariant for the full group. On the other hand, by Corollary 24.8, we are entitled to a $SL_2(\mathbf{F}_p)$-equivariant flag for $\mathcal{E}$ or, what is the same in this case, a $SL_2(\mathbf{F}_p)$-invariant line subbundle of $\mathcal{E}$.

Lemma (24.15): *The center of $SL_2(\mathbf{F}_p)$ acts trivially on the vector bundle $\mathcal{E}$ if and only if p is congruent to 1 modulo 4.*

Proof: The vector bundle $\mathcal{E}$ is a subbundle of the trivial V^+ bundle on $X(p)$. It is well known (and follows from the results of Appendix I) that the center of $SL_2(\mathbf{F}_p)$ acts trivially on V^+ if and only if $p \equiv 1 \pmod 4$.

Let us now try to say something about $SL_2(\mathbf{F}_p)$-invariant rank 2 bundles on $X(p)$ in general. Let E be such a bundle. We know from Theorem 24.1 and Corollary 24.4 that every $SL_2(\mathbf{F}_p)$-invariant line bundle on $X(p)$ is of the form λ^n for some $n \in \mathbf{Z}$. Therefore, from the existence of a $SL_2(\mathbf{F}_p)$-invariant flag on E, we obtain an exact sequence of the form

$$0 \to \lambda^m \to E \to \lambda^n \to 0.$$

The class of the extension of λ^n by λ^m defined by this exact sequence lies in the cohomology group

$$H^1(X(p), HOM(\lambda^n, \lambda^m)) = H^1(X(p), \lambda^{m-n}).$$

Furthermore, since the flag is $SL_2(\mathbf{F}_p)$-invariant, the class must be fixed by $SL_2(\mathbf{F}_p)$. Therefore, such extensions are described by a triple (m, n, c) where m, n are integers and c is an element of

$$H^0(SL_2(\mathbf{F}_p), H^1(X(p), \lambda^{m-n})).$$

We are thus led to study the multiplicity of the trivial representation in the $SL_2(\mathbf{F}_p)$-module $H^1(X(p), \lambda^{m-n})$. The main tool we use for this purpose is the Woods Hole Fixed Point Formula. However, without this formula, we can prove the following result.

Lemma (24.16): *Let m, n be integers. If $m - n$ is odd then every $SL_2(\mathbf{F}_p)$-equivariant exact sequence of vector bundles of the form*

$$0 \to \lambda^m \to E \to \lambda^n \to 0$$

splits. In particular, any $SL_2(\mathbf{F}_p)$-invariant rank 2 vector bundle on $X(p)$ of odd degree is a direct sum of $SL_2(\mathbf{F}_p)$-invariant line bundles.

Proof: Replacing E by its dual if necessary, we can assume that m is even. Denote by $1, \epsilon$ the trivial and the nontrivial characters of the center of $SL_2(\mathbf{F}_p)$. If ψ is one of them, then $E(\psi)$ is a $SL_2(\mathbf{F}_p)$-invariant subbundle of E. Since m is even, n is odd and we have $\lambda^m = \lambda^m(1)$ and $\lambda^n = \lambda^n(\epsilon)$. Therefore, one of the bundles $\lambda^m(\psi)$, $\lambda^n(\psi)$ must be 0. It follows that $E(\psi)$ cannot be all of E. Therefore, E is the direct sum of $\lambda^m = E(1)$ and $E(\epsilon)$, the latter being mapped isomorphically onto λ^n. Thus the sequence splits and E is a direct sum of line bundles.

It follows from this result that we can confine our attention to invariant rank 2 bundles of even degree. For such a vector bundle, we have an equivariant exact sequence of invariant vector bundles

$$0 \to \lambda^m \to E \to \lambda^n \to 0$$

in which m, n have the same parity. By replacing E by $E \otimes \lambda^{-n}$ if necessary, we can assume that m is even and $n = 0$. In this case, to which we confine our attention, the center of $SL_2(\mathbf{F}_p)$ acts trivially on E and it is actually the group $PSL_2(\mathbf{F}_p)$ that acts on E.

For an integer ν, denote by χ_ν the Euler character of $PSL_2(\mathbf{F}_p)$ acting on the cohomology of λ^ν, i.e the alternating sum of the characters of $PSL_2(\mathbf{F}_p)$ acting on the cohomology spaces of λ^ν. If γ is an element of $PSL_2(\mathbf{F}_p)$ acting nontrivially on $X(p)$, i.e. an element other than the identity element, we have

$$(24.17) \qquad \chi_\nu(\gamma) = tr(\gamma \mid H^0(X(p), \lambda^\nu) - tr(\gamma \mid H^1(X(p), \lambda^\nu) = \sum_{\gamma(x) = x} \frac{\lambda_x^\nu(\gamma)}{1 - d\gamma_x},$$

where $\lambda_x^\nu(\gamma)$ denotes the scalar multiplication by which γ acts on the fibre λ_x^ν of λ^ν at x and $d\gamma_x$ the scalar by which γ acts on the fibre κ_x of the canonical bundle at x. We will denote by ζ_p a primitive p-th root of unity. The expression ζ_p^a is then meaningful for any integer or for any element of the field $\mathbf{F}_p$. If r is a rational number whose denominator is not divisible by p, then the expression ζ_p^r uniquely defines a p-th root of unity, namely ζ_p^a, where a is the residue class of r modulo p. We will denote by Tr the trace from the cyclotomic field $Q(\zeta_p)$ to its unique quadratic subfield.

Lemma (24.18):   *Let ν be an integer. Then for $\gamma \in PSL_2(\mathbf{F}_p)$ we have*

$$
\chi_\nu(\gamma) = \begin{cases}
0 & \text{if } \gamma \text{ does not have order 1,2,3 or } p; \\[2mm]
Tr\left(\dfrac{\zeta_p^{-\nu t/12}}{1-\zeta^t}\right) & \text{if } \gamma \text{ is conjugate to } \begin{pmatrix} 1 & t \\ 0 & 1 \end{pmatrix}; \\[3mm]
\dfrac{(1-\eta_3)(p-\epsilon_3)}{6} & \text{if } \gamma \text{ has order 3}; \\[3mm]
\dfrac{p-\epsilon_2}{4}(\epsilon i)^\nu & \text{if } \gamma \text{ has order 2}; \\[3mm]
\dfrac{(\nu-p+6)(p^2-1)}{24} & \text{if } \gamma = 1.
\end{cases}
$$

Here $\epsilon = \pm 1$ independently of γ when γ has order 2; ϵ_2, ϵ_3 are the quadratic characters of -1, -3 respectively modulo p; and η_3 is the unique integer among $0,1,2$ which is congruent to $p\nu$ modulo 3.

Proof: If γ does not have order 1,2,3 or p, then γ has no fixed points and the summation on the right hand side of the fixed point formula is empty, hence 0. If γ is the identity element of $PSL_2(\mathbf{F}_p)$, then the value of $\chi_n u$ is given by the Riemann-Roch theorem:

$$
\begin{aligned}
\chi_\nu(1) &= h^0(\lambda^\nu) - h^1(\lambda^\nu) = deg(\lambda^\nu) + 1 - g \\
&= \frac{p^2-1}{24}\nu - \frac{(p-6)(p^2-1)}{24} \\
&= \frac{(\nu-p+6)(p^2-1)}{24},
\end{aligned}
$$

where g denotes the genus of the modular curve. If γ has order p then it is conjugate to an element of the form $\begin{pmatrix} 1 & t \\ 0 & 1 \end{pmatrix}$, with $t \neq 0$, and the conjugacy class of γ depends only on the quadratic residue symbol of t. In either case, γ will have exactly $(p-1)/2$ fixed points and the eigenvalues $d\gamma_x$ as x runs over these fixed points will be the values ζ_p^{tb}, where b runs over the quadratic residues modulo p. On the other hand, we have $\kappa = \lambda^{2p-12}$ by Cor.24.4 and since -12 is relatively prime to p, we therefore have

$$
\lambda_x(\gamma) = \zeta_p^{-tb/12}
$$

and

$$
\lambda_x^\nu(\gamma) = \zeta_p^{-\nu tb/12}.
$$

Therefore the right hand side of the fixed point formula becomes

$$
{\sum}' \frac{\zeta_p^{-\nu tb/12}}{1-\zeta_p^{tb}} = Tr\left(\frac{\zeta_p^{-\nu t/12}}{1-\zeta_p^t}\right).
$$

Let ω be a primitive cube root of unity. If γ has order 3 then it has $(p-\epsilon_3)/3$ fixed points, where $\epsilon_3 = \pm 1$ is congruent to p modulo 6, i.e. ϵ_3 is the quadratic residue symbol of -3 modulo p. At half of these fixed points, the value of $d\gamma_x$ will be ω and at the other half it will be ω^2. Since $\kappa = \lambda^{2p-12}$, we therefore have

$$\lambda_x^\nu(\gamma) = d\gamma_x^{2p\nu}$$

and the right side of the fixed point formula becomes

$$\frac{p - \epsilon_3}{6} \left(\frac{\omega^{2p\nu}}{1 - \omega} + \frac{\omega^{p\nu}}{1 - \omega^2} \right) = \left\{ \begin{array}{ll} 1 & \text{if } p\nu \equiv 0 \pmod 3 \\ 0 & \text{if } p\nu \equiv 1 \pmod 3 \\ -1 & \text{if } p\nu \equiv 2 \pmod 3 \end{array} \right\} = \frac{(1 - \eta_3)(p - \epsilon_3)}{6}.$$

If γ has order 2 then it has $(p - \epsilon_2)/2$ fixed points, where $\epsilon_2 = \pm 1$ is congruent to p modulo 4, i.e. ϵ_2 is the quadratic residue symbol of -1 modulo p. At all of these fixed points, we have $d\gamma_x = -1$, from which it follows that

$$\lambda_x(\gamma) = \pm i.$$

Furthermore, the sign of i will be the same at all of the fixed points of γ. Denoting the sign by ϵ, we then have

$$\lambda_x^\nu(\gamma) = (\epsilon i)^\nu,$$

so we only need to worry about the sign ϵ when ν is odd. This can be done by a more careful analysis of the representation $\rho_t^\pm$ and the geometry of $X(p)$ then we need to go into here. The value of $\chi_\nu(\gamma)$ is then

$$\frac{p - \epsilon_2}{4} (\epsilon i)^\nu.$$

Lemma (24.19): *Let $\nu = 2\mu$ be an even integer. Let ϵ, ϵ_2, ϵ_3 and η_3 be as in the preceding lemma. Let the integers a, τ be determined by the conditions that $0 \le a < p$ and $12a = \nu + \tau p$. Then the multiplicity $(1, \chi_\nu)$ of the trivial character in the virtual character χ_ν is given by*

$$(1, \chi_\nu) = -\frac{4\eta_3 + \tau - 3(-1)^\mu + 3}{12}.$$

Proof: The multiplicity is given by

$$\frac{1}{|PSL_2(\mathbf{F}_p)|} \sum \chi_\nu(\gamma)$$

where the summation runs over all elements γ in $PSL_2(\mathbf{F}_p)$. Let t denote a quadratic nonresidue modulo p. Let $\mathcal{T}r$ denote the trace from the cyclotomic field $\mathbf{Q}(\zeta_p)$ to the field of rational numbers. Taking just the summation and grouping its terms according to the conjugacy class of γ, we obtain

$$1 \cdot \frac{(\nu - p + 6)(p^2 - 1)}{24} + \frac{p^2 - 1}{2} \cdot Tr\left(\frac{\zeta_p^a}{1 - \zeta_p}\right)$$

$$+ \frac{p^2 - 1}{2} \cdot Tr\left(\frac{\zeta_p^{at}}{1 - \zeta_p^t}\right) + 2 \cdot \frac{p^3 - p}{p - \epsilon_3} \cdot (1 - \eta_3)\frac{p - \epsilon_3}{6}$$

$$+ \frac{p^3 - p}{p - \epsilon_2} \cdot \frac{p - \epsilon_2}{4}(\epsilon i)^\nu$$

$$= 1 \cdot \frac{(\nu - p + 6)(p^2 - 1)}{24} + \frac{p^2 - 1}{2} \cdot Tr\left(\frac{\zeta_p^a}{1 - \zeta_p}\right)$$

$$+ 2 \cdot \frac{p^3 - p}{6} \cdot (1 - \eta_3) + \frac{p^3 - p}{4} \cdot (-1)^\mu$$

The following identity is easily verified by multiplying both sides by $1 - \zeta_p$:

$$\frac{1}{1 - \zeta_p} = -\frac{1}{p}(1 + 2\zeta_p + 3\zeta_p^2 + \ldots + p\zeta_p^{p-1}).$$

Using it we can easily compute the Tr term as follows.

$$Tr\frac{\zeta_p^a}{1 - \zeta_p} = -\frac{1}{p}Tr((1 + 2\zeta_p + 3\zeta_p^2 + \ldots + p\zeta_p^{p-1})\zeta_p^a)$$

$$= -\frac{1}{p}Tr(\zeta_p^a + 2\zeta_p^{a+1} + \ldots + p\zeta_p^{a+p-1})$$

$$= -\frac{1}{p}(-1 - 2 - 3 - \ldots - p + p(p - a + 1))$$

$$= a - \frac{p+1}{2}.$$

Therefore the multiplicity of the trivial representation in χ_ν becomes

$$\frac{2}{p^3 - p} \cdot \frac{p^2 - 1}{24}[(\nu - p + 6) + 12(a - \frac{p+1}{2}) + 4p(1 - \eta_3) + 3p(-1)^\mu]$$

$$= \frac{1}{12p}[\nu - p + 6 - \nu - p\tau - 6p - 6 + 4p(1 - \eta_3) + 3p(-1)^\mu]$$

$$= \frac{1}{12}[-7 - \tau + 4(1 - \eta_3) + 3(-1)^\mu]$$

$$= -\frac{4\eta_3 + \tau - 3(-1)^\mu + 3}{12}.$$

The reader might find it instructive and reassuring to verify directly that $4\eta_3 + \tau - 3(-1)^\mu + 3$ is indeed a multiple of 12.

In our study of extensions of λ^n by λ^m, it is not χ_ν, with $\nu = m - n$, that we want, but the character $SL_2(\mathbf{F}_p)$ on $H^1(X(p), \lambda^\nu)$. However, in case ν is negative, $H^0(X(p), \lambda^\nu) = 0$ and the character on $H^1(X(p), \lambda^\nu)$ is in fact given by $-\chi_\nu$. Therefore we have the following result.

Corollary (24.20):   If ν is a negative even integer, then the multiplicity of the trivial representation of $PSL_2(\mathbf{F}_p)$ on $H^1(X(p), \lambda^\nu)$ is equal to

$$\frac{4\eta_3 + \tau - 3(-1)^\mu + 3}{12}.$$

The special case $\nu = 3 - p$ is of interest in connection with the vector bundle $\mathcal{E}$.

Corollary (24.21):   Let $\nu = 3 - p$. Then the multiplicity of the trivial representation of $PSL_2(\mathbf{F}_p)$ on $H^1(X(p), \lambda^\nu)$ is equal to

(24.22)
$$\frac{1 + \epsilon_2}{2}.$$

Thus, it is 1 if p is 1 modulo 4 and 0 if p is 3 modulo 4. In particular, there is no nonsplit $PSL_2(\mathbf{F}_p)$-equivariant extension of ζ^{-2} by 1 for p congruent to 3 modulo 4 and there is an essentially unique nonsplit equivariant extension if p is congruent to 1 modulo 4.

Proof: Since $\nu = 3 - p$, we have $p\nu = 3p - p^2$ congruent to 2 modulo 3. Therefore $\eta_3 = 2$. Also,

$$(-1)^\mu = (-1)^{(3-p)/2} = -\epsilon_2.$$

To compute τ, we consider p modulo 4. For example, if $p = 4n + 1$ then modulo p we have

$$a \equiv -\frac{\nu}{12} \equiv \frac{p-3}{12} \equiv \frac{3p-3}{12} = \frac{p-1}{4},$$

so we can take $a = (p-1)/4 = n$ and

$$\tau = \frac{-12a - \nu}{p} = \frac{3 - 3p - (3 - p)}{p} = -2.$$

Similarly, if $p = 4n + 3$, then modulo p we have

$$a \equiv -\frac{\nu}{12} \equiv \frac{p-3}{12} \equiv \frac{9p-3}{12} = \frac{3p-1}{4},$$

so we can take $a = (3p-1)/4 = 3n + 2$ and

$$\tau = \frac{-12a - \nu}{p} = \frac{3 - 9p - (3 - p)}{p} = -8.$$

Therefore

$$\tau = \begin{cases} -2 & \text{if } \epsilon_2 = 1, \\ -8 & \text{if } \epsilon_2 = -1 \end{cases} = -5 + 3\epsilon_2.$$

The multiplicity of the trivial representation in $H^1(X(p), \lambda^\nu)$ is therefore equal to

$$\frac{4\eta_3 + \tau - 3(-1)^\mu + 3}{12} = \frac{8 - 5 + 3\epsilon_2 + 3\epsilon_2 + 3}{12} = \frac{1 + \epsilon_2}{2}.$$

Since we know that the exact sequence

$$0 \to \zeta^{-2} \to \mathcal{E} \to 1 \to 0$$

does not split, we conclude that not only is the given embedding of ζ^{-2} not equivariant but there is no equivariant embedding in case p is congruent to 3 modulo 4.

We believe that in general the dimensions of $H^0(X(p), \zeta)$ and $H^0(X(p), \alpha)$ are $(p-1)/2$ and $(p+1)/2$ respectively, with related results for low powers of α and ζ, but we have not proved this. We will call this the **WYSIWYG Hypothesis**.[15]

In the case $p = 11$, the first author has found other $SL_2(\mathbf{F}_p)$ invariant vector bundles of rank 2 on $X(p)$. See the articles [A15], [A16] for details and generalizations.

---

[15] Readers familiar with certain types of text editors will probably recognize the acronym "WYSYWYG" as standing for "What you see is what you get."

Appendix I

On the Weil representation
by Allan Adler

In his fundamental papers in Acta Mathematica, A. Weil [W1],[W4] introduced the study of the group $\mathbf{B}_0(G)$, which is a group of unitary operators which is also an extension by a circle group of a symplectic group over a locally compact abelian group G. By considering splittings of this extension over various subgroups of the symplectic group, one obtains unitary representations of these subgroups. At the same time, the extension itself can be viewed as defining a projective unitary representation of the symplectic group. All of these representations, projective or otherwise, are commonly referred to in various contexts as "the" Weil representation and various subgroups of $\mathbf{B}_0(G)$ for special choices of G are commonly referred to as "the" metaplectic group.

In case G is a finite abelian group of odd order, it is well known that the extension $\mathbf{B}_0(G)$ splits and that the Weil representation of $Sp(G)$ is actually a representation, not just a projective representation. In [W1], Weil wrote down explicit transformations for the projective unitary representation. The problem of giving explicit transformations for the lifting of the projective representation in the case of the finite symplectic group $Sp(G)$ and various subgroups of Lie type was solved in general by Gérardin [Gé]. Some earlier work was done by Tanaka [T1],[T2],[T3],[T4] in the case of $SL(2, \mathbf{Z}/p^\lambda \mathbf{Z})$ and $SL_2(F)$ where F is a finite field of odd characteristic. We prefer to retain the notation of [We1], which is not done in [Gé], in order to give an explicit lifting of Weil's projective representation of a finite symplectic group $Sp_{2n}(\mathbf{F}_p)$. The key idea, however, appears to be well known independently of [Gé], judging by a remark in the paper [F-T] of Feit and Tits, where it is described as "well known and obvious". The author has continued the study of the topics of this appendix in an appendix to the article [A6].

§25   Generalization of the Metaplectic Group

Let $\mathbf{T}$ denote the multiplicative group of all complex numbers of absolute value 1. Let G be a locally compact abelian group such that $x \mapsto 2x$ is an automorphism of G. Denote by $L^2(G)$ the space of all square integrable functions with respect to a Haar measure dg on G. Let G^* denote the dual group of G, i.e. the group of all continuous homomorphisms from G to $\mathbf{T}$, with the topology of uniform convergence on compact sets; then G^* is likewise a locally compact abelian group and the dual group of G^* is canonically isomorphic to G. There is a natural continuous pairing

$[\cdot, \cdot]$ from $G \times G^*$ to $\mathbf{T}$ given by

$$[g, g^*] = g^*(g)$$

for all $(g, g^*) \in G \times G^*$. Let $\mathbf{A}(G)$ be the group of all operators on $L^2(G)$ of the form $U(g, g^*, t)$, for $g \in G$, $g^* \in G^*$ and $t \in T$, given by

$$U(g, g^*, t)\Phi(x) = t[x, g^*]\Phi(g + x)$$

for all $x \in G$. The normalizer of $\mathbf{A}(G)$ in the group of all unitary operators on $L^2(G)$ is denoted $\mathbf{B}_0(G)$. On the other hand, we denote by $B_0(G)$ the group of all continuous automorphisms of $\mathbf{A}(G)$ which induce the identity automorphism on the center of $\mathbf{A}(G)$. There is a natural homomorphism from $\mathbf{B}_0(G)$ to $B_0(G)$ which associates to an element $U \in \mathbf{B}_0(G)$ the automorphism $V \mapsto UVU^{-1}$ of $\mathbf{A}(G)$. According to Theorem 1 of [W1], this homomorphism from $\mathbf{B}_0(G)$ to $B_0(G)$ is surjective with kernel $\mathbf{T}$, where $\mathbf{T}$ is identified with the group of unitary scalar multiplications on $L^2(G)$.

We denote by $Sp'(G)$ the subgroup of $B_0(G)$ consisting of all s in $B_0(G)$ which commute with $d_0(-1_G)$. The following lemma is an easy consequence of §5 of [W1].

Lemma (25.1): *An element $s = (\sigma, f)$ of $B_0(G)$ lies in $Sp'(G)$ if and only if $f(w) = f(-w)$ for all w in $G \times G^*$. Furthermore, if σ' is any symplectic automorphism of $G \times G^*$ there is one and only one second degree character f' of $G \times G^*$ such that $s' = (\sigma', f')$ belongs to $Sp'(G)$. Hence, the natural homomorphism $(\sigma, f) \mapsto \sigma$ of $B_0(G)$ onto $Sp(G)$ maps $Sp'(G)$ isomorphically onto $Sp(G)$.*

We denote by (E) the following hypothesis:

(E)   The group $B_0(G)$ is generated by $\Omega_0(G)$.

We will denote by $\mathbf{B}_1(G)$ the subgroup of $\mathbf{B}_0(G)$ consisting of all operators S in $\mathbf{B}_0(G)$ which commute with the operator $\mathbf{d}_0(-1_G)$.

Lemma (25.2): *The canonical projection π_0 maps $\mathbf{B}_1(G)$ onto a subgroup of $Sp(G)$ containing $Sp'(G) \cap \Omega_0(G)$. If Hypothesis (E) holds then π_0 maps $\mathbf{B}_1(G)$ surjectively onto $Sp'(G)$.*

Proof: This follows easily from Proposition 1 of §7 of [W1], the explicit formulas used in its proof, the formulas

$$\mathbf{d}_0'(\gamma\alpha) = \mathbf{d}_0'(\gamma)\mathbf{d}'(\alpha)$$

$$\mathbf{d}_0(\alpha)^{-1}\mathbf{t}_0(f)\mathbf{d}_0(\alpha) = \mathbf{t}_0(f^\alpha)$$

of §13 of [W1] and the fact that

$$\pi \circ \mathbf{d}_0 = d_0$$
$$\pi \circ \mathbf{d}_0' = d_0'$$
$$\pi \circ \mathbf{t}_0 = t_0.$$

Details are left to the reader.

Remark (25.3): In case G is a free module of finite rank over a local field or over the ring of adèles of an **A**-field, the metaplectic group is defined (cf. §§34,37 of [W1]) and is canonically isomorphic to $\mathbf{B}_1(G)$. Hence, for an arbitrary locally compact abelian group such that $x \mapsto 2x$ is an automorphism of G, we will call $\mathbf{B}_1(G)$ the **metaplectic group** of G.

§26   The Weil Representation of Finite Symplectic Groups

Denote by V^+ and V^- the subspaces of $L^2(G)$ consisting of even functions and odd functions respectively. The V^+ and V^- are both invariant under the metaplectic group $\mathbf{B}_1(G)$. For every S in $\mathbf{B}_1(G)$ we will write S^+ and S^- to denote the operators induced on V^+ and V^- respectively by S.

From now on, G will be a finite abelian group of odd order $2N + 1$. We then have a simple lemma whose proof we leave to the reader.

Lemma (26.1): *Define the complex-valued function χ on $\mathbf{B}_1(G)$ by the rule*

$$\chi(S) = \frac{det(S^+)}{det(S^-)}$$

where det denotes determinant. Then χ is a character of the group $\mathbf{B}_1(G)$. Furthermore, if t is a complex number of absolute value 1 and if $\mathbf{t}$ is the scalar multiplication by t on $L^2(G)$, then we have

$$\chi(\mathbf{t}) = t.$$

It follows from Lemma 26.1 that the mapping φ from $\mathbf{B}_1(G)$ to itself given by

$$\varphi(S) = \chi(S)^{-1}S$$

is a homomorphism. Denote by $Sp''(G)$ the image of the homomorphism φ. It is easy to see that $\mathbf{B}_1(G)$ is the direct product of $Sp''(G)$ and the group **T**. We denote by $B_1(G)$ the image of $\mathbf{B}_1(G)$ under the canonical projection π_0. Then since **T** is the kernel of π_0, it follows that the canonical projection maps $Sp''(G)$ isomorphically onto $B_1(G)$. Hence we have the following consequence of Lemma 25.2.

Lemma (26.2): *If G is a finite abelian group of odd order satisfying hypothesis (E) then the canonical projection π_0 maps $Sp''(G)$ isomorphically onto $Sp'(G)$.*

Lemma 26.2 gives us an isomorphism of $Sp''(G)$ onto $Sp'(G)$ whose inverse we will denote by $\mathbf{r}'$ in imitation of Weil's notation $\mathbf{r}_r$. The isomorphism $\mathbf{r}'$ can be viewed as a unitary representation of $Sp'(G)$ on $L^2(G)$. We will refer to $\mathbf{r}'$ as the **Weil representation** of $Sp'(G)$ on $L^2(G)$. Our task in the next section will be to describe $\mathbf{r}'$ explicitly in certain cases.

§27   Explicit Form of the Weil Representation

Let R be full set of representatives for the orbits of the automorphism -1_G if G. Let β^+ denote the set of all functions of the form $\delta_g + \delta_{-g}$ on G with g in G and let β^- denote the set of all functions of the form $\delta_g - \delta_{-g}$ with $g \neq 0$ in R. Here δ_x denotes the function which is 1 at x and which is 0 elsewhere on G. Then β^+ and β^- are bases for V^+ and V^- respectively.

Theorem (27.1):   *Let f be a second degree character of G such that $f(x) = f(-x)$ for all x in G. Then $t_0(f)$ belongs to $B_1(G)$ and we have*

$$r'(t_0(f)) = t_0(f).$$

Proof: The first assertion follows from §13 of [W1], in particular from the identity

$$t_0(f) = \pi_0(t_0(f)).$$

Let $S = t_0(f)$. Then we have $r'(t_0(f)) = \chi(S)^{-1}S$ and we have to prove that $\chi(S) = 1$. If g is any element of R we have

$$S(\delta_g + \epsilon\delta_{-g}) = f(g) \cdot (\delta_g + \epsilon\delta_{-g})$$

for $\epsilon = \pm 1$ and we have $f(0) = 1$. Therefore we have

$$det(S^+) = det(S^-) = \prod f(g)$$

where the product runs over all g in R. So $\chi(S) = 1$.

From now on we will assume that G is a vector space of finite dimension n over the field $\mathbf{F}_p$ where p is an odd prime. In this case, it is easy to prove that hypothesis (E) holds. Therefore the conclusion of Lemma 26.4 of §2 holds as well.

Theorem (27.2):   *Let ρ be a nondegenerate symmetric isomorphism of G onto G^*. Then we have*

$$\chi(d_0'(\rho^{-1})) = \gamma(f)$$

where $\gamma(f)$ is as defined in Theorem 2 of §14 of [W1] and where f is the nondegenerate second degree character of G associated to ρ such that $f(x) = f(-x)$ for all x in G.

Proof: Let $\gamma = \gamma' = 2\rho^{-1}$ and let $\gamma'' = 4\rho^{-1}$. Then we have

$$t_0'(f') \cdot t_0'(f') = t_0'(f'')$$

where f' and f'' are nondegerate second degree characters of G^* associated to γ and γ'' respectively and which satisfy

$$f'(-x^*) = f'(x^*)$$
$$f''(-x^*) = f''(x^*)$$

for all x^* in G^*. Since $t_0'(f')$ and $t_0'(f'')$ lie in $\Omega_0(G)$ we have by Theorem 3 of §15 of [W1] that

$$\mathbf{r}_0(s)\mathbf{r}_0(s') = \gamma(f)\mathbf{r}_0(s'').$$

If we apply χ to both sides of this identity then by Theorem 27.1 of this paper and by Proposition 1 of [W1] and the formulas appearing in its proof we have

$$\chi(\mathbf{d}_0'(\gamma))^2 = \gamma(f)\chi(\mathbf{d}_0'(\gamma'')).$$

Since $\gamma'' = 2\gamma$ we have

$$\mathbf{d}_0'(\gamma'') = \mathbf{d}_0'(\gamma)\mathbf{d}_0(2 \cdot 1_G).$$

Therefore

$$\chi(\mathbf{d}_0'(2\rho^{-1})) = \gamma(f)\chi(\mathbf{d}_0(2 \cdot 1_G))$$

and hence

$$\gamma(f) = \chi\left(\mathbf{d}_0'(2\rho^{-1})\mathbf{d}_0(\frac{1}{2} \cdot 1_G)\right) = \chi(\mathbf{d}_0'(\rho^{-1})).$$

Corollary (27.3): *Let ρ and f be as in Theorem 27.2. Then*

$$\chi(\mathbf{d}_0'(\rho^{-1})) = \int_G f \, d\mu$$

where the integral is taken with respect to the self-dual Haar measure on G.

Proof: By Theorem 2 of §14 of [W1], we have

$$\mathcal{F}(f)(x^*) = \gamma(f)|\rho|^{-\frac{1}{2}}f(x^*\rho^{-1})^{-1}$$

where $\mathcal{F}(f)$ denotes the Fourier transform of the second degree character f. If we evaluate both sides of this identity at the identity element of G^* we get

$$\gamma(f) = \int_G f \, d\mu$$

where the integral is taken with respect to the self-dual Haar measure on G. The result now follows from Theorem 27.2.

Theorem (27.4): *Let α be an automorphism of G. Then we have*

$$\mathbf{r}'(\mathbf{d}_0(\alpha)) = \left(\frac{det(\alpha)}{p}\right)\mathbf{d}_0(\alpha).$$

Proof: Define $\nu : Aut(G) \to \mathbf{C}^\times$ by

$$\nu(\alpha) = \chi(\mathbf{d}_0(\alpha)).$$

Then ν is a character. Therefore if α lies in the commutator subgroup of $Aut(G)$ we have $\nu(\alpha) = 1$. This shows that $\nu(\alpha)$ depends only on the determinant of α. Therefore we may assume that α is a represented by a diagonal matrix whose diagonal entries are $\Delta, 1, 1, \ldots, 1$. We can chose R such that the first nonzero coordinate of any element of R lies between 1 and $\frac{1}{2}(p-1)$ and this determines R completely. Now, $\mathbf{d}_0(\alpha)$ fixes $\delta_g \pm \delta_{-g}$ for all g in G whose first coordinate vanishes. So let R^*

denote the subset of R consisting of all g in R whose first coordinate is nonzero. Write $G = \mathbf{F}_p \oplus G'$ where G' consists of all vectors in G whose first coordinate vanishes and where $\mathbf{F}_p$ is identified with the space of all vectors whose remaining coordinates vanish. Let π denote the projection of G onto G'. For each g' in G', the action of α on $\pi^{-1}(g')$ is isomorphic to the action of Δ by multiplication of $\mathbf{F}_p$. If the theorem is true in the case $G = \mathbf{F}_p$ then we will have

$$\chi(\mathbf{d}_0(\alpha)) = \left(\frac{\Delta}{p}\right)^{\circ(G')} = \left(\frac{\Delta}{p}\right)$$

since G' has odd order. Therefore we may assume $n = 1$. But then the theorem reduces to the theorem of Gauss that

$$\left(\frac{\Delta}{p}\right) = (-1)^N$$

where N is the number of elements k of $\mathbf{F}_p$ such that $1 \leq k \leq \frac{1}{2}(p-1)$ and such that $\Delta k = l$ with $\frac{1}{2}(p+1) \leq l \leq p-1$.

Theorem (27.5): *Let $\gamma : G^* \to G$ be any isomorphism. Let ρ be a symmetric isomorphism of G onto G^* and let α be the automorphism of G defined by $\gamma = \rho^{-1}\alpha$. Then*

$$\mathbf{d}_0'(\gamma) = \mathbf{d}_0'(\rho^{-1})\mathbf{d}(\alpha)$$

and we have

$$\chi(\mathbf{d}_0'(\gamma)) = \gamma(f)\left(\frac{det(\alpha)}{p}\right)$$

where f is the nondegenerate second degree character of G associated to ρ such that $f(x) = f(-x)$ for all x in G.

Proof: This follows at once from Theorems 27.2 and 27.4.

Theorem 27.4, Theorem 27.5 and Corollary 27.3 amount to an explicit determination of the Weil representation $\mathbf{r}'$. The appendix of [A6] is devoted to a detailed study of fundamental intertwining operator of [A-R], Lemma 20.21 from the point of view adopted here.

Appendix II

Modular Forms of Weight 4/5 for $\Gamma(11)$
by Allan Adler

§28   Introduction

In this paper, we show how some old and neglected geometrical studies of modular forms by Felix Klein [K1], recently resurrected by W.L.Edge [E1] and in turn by the author [A1-11] allow us to construct some explicit modular forms of weight 4/5 and level 11. In the first section, we recall some of the author's results on the ring of invariants of the irreducible 5 dimensional representation of $PSL_2(\mathbf{F}_{11})$ and introduce the notion of a standard framing of such a representation. In the second section, we recall Klein's remarkable result that the modular curve of level 11 is isomorphic to the singular locus of the unique quintic invariant.[16] Using the determination, due to the author and S.Ramanan (cf. §24) of the group of all invariant line bundles on a modular curve of prime level, we show how Klein's theorem implies that a natural space of modular forms of weight 4/5 arises from the relevant geometry. In the final section, we show that the geometry and the invariant theory allow us to compute these modular forms of weight 4/5 explicitly and painlessly.

The author first suggested in [A2] that one might be able to use Klein's theorem to construct modular forms of weight 4/5. However, it was only years later, while visiting the University of Rhode Island in Kingston recently, that the author finally saw the simple method of doing so.

§29   Representations and Invariants of $\mathbf{PSL_2(F_{11})}$

If f, g are polynomials in the variables $x_1, \ldots, x_n$, we will denote by $f \# g$ the expression

$$f(\frac{\partial}{\partial x_1}, \ldots, \frac{\partial}{\partial x_n})g.$$

---

[16] Current work of the author suggests that this and similar loci are interesting from the standpoint of mirror symmetry. This will be discussed further in [A15] and [A16].

In other words, we compute $f\#g$ by replacing each of the variables of f by the corresponding partial derivative operator and let the resulting differential operator act on g. The proof of the following simple lemma is left to the reader.

Lemma (29.1): *The operation* $\#$ *is bilinear. Furthermore, if* T *is a linear automorphism of the span of* $x_1, \ldots, x_n$ *and if we extend it to an automorphism, also denoted* T, *of the polynomial ring* $\mathbf{C}[x_1, \ldots, x_n]$, *then we have*

$$T(f\#g) = ({}^tT^{-1}f)\#(T(g)),$$

where the extension of ${}^tT^{-1}$ *to an automorphism of the polynomial ring is also denoted* ${}^tT^{-1}$.

We will also allow f or g to be vectors whose entries are polynomials.

In my paper [A9], I have computed the ring of invariants for the action of $PSL_2(\mathbf{F}_{11})$ in an irreducible complex representation of degree 5. The ring is generated by 10 invariants

$$f_3 \ f_5 \ f_6 \ f_7 \ f_8 \ f_9 \ f_{10} \ f_{11} \ f_{12} \ f_{14},$$

where f_i has degree i. The invariants $f_3, f_5, f_6, f_8, f_{11}$ are a homogeneous system of parameters and the full ring of invariants is a free module of rank 12 over $\mathbf{C}[f_3, f_5, f_6, f_8, f_{11}]$ with basis

$$1 \ \ f_7 \ \ f_9 \ \ f_{10} \ \ f_{12} \ \ f_{14} \ \ f_7^2 \ \ f_7 f_9 \ \ f_9^2 \ \ f_9 f_{10} \ \ f_7^3 \ \ f_9^2 f_{10}.$$

The reader is referred to [A9] for details of the definition and geometric interpretation of the invariants. Here we will be concerned with three of the invariants: f_3, f_5, f_7. The invariant f_3 is the cubic form

$$f_3 = v^2w + w^2x + x^2y + y^2z + z^2v,$$

which was studied by Felix Klein [K2], [K-F] and by Francesco Brioschi [Bri]. The invariant f_5 is, up to a trivial factor of 2, the Hessian of f_3, i.e. the determinant of the matrix of second partials of f_3. That matrix is exhibited in Theorem 30.1 below. Hence, f_5 is given by

$$f_5 = 3vwxyz + \sigma(v^3x^2 - v^3yz),$$

where σ denotes summation over the powers of the cyclic permutation $(vwxyz)$. For example, $f_3 = \sigma(v^2w)$. The invariant f_7 is a little more complicated, being given by

$$f_7 = \sigma(v^6z + 3v^5y^2 - 15v^4wxz + 5v^3w^3y + 15v^3wxy^2),$$

or more concisely by

$$f_7 = \frac{1}{160}D^2(f_3f_5^2)$$

where D is the differential operator defined by

$$D(g) = f_3\#(f_3g).$$

Just as the Hessian may be interpreted as the locus of points whose polar quadrics[17] with respect to the cubic f_3 are cones, the invariant f_7 may be interpreted geometrically as the locus of all points p such that the polar quadric of p with respect to the the Hessian is outpolar to the polar quadric of p with respect to the cubic.

Denote by P and R the following matrices:

$$P = \begin{pmatrix} \zeta & 0 & 0 & 0 & 0 \\ 0 & \zeta^9 & 0 & 0 & 0 \\ 0 & 0 & \zeta^4 & 0 & 0 \\ 0 & 0 & 0 & \zeta^3 & 0 \\ 0 & 0 & 0 & 0 & \zeta^5 \end{pmatrix}$$

$$R = \begin{pmatrix} 0 & 0 & 0 & 0 & 1 \\ 1 & 0 & 0 & 0 & 0 \\ 0 & 1 & 0 & 0 & 0 \\ 0 & 0 & 1 & 0 & 0 \\ 0 & 0 & 0 & 1 & 0 \end{pmatrix}$$

Here $\zeta = exp(2\pi i/11)$. It is easy to see that P and R preserve f_3, f_5, f_7 and are unitary. The matrix (cf. [A7], [K3])

$$J = -\frac{1}{\sqrt{-11}} \sum_{i=1}^{5} R^i (P^2 - P^{-2}) R^i$$

also preserves these forms and has order 2. It is also unitary. Together, P and J generate the full group of order 660, which we denote by G. The natural action of this group of matrices on $\mathbf{C}^5$ is denoted ρ. The dual of the representation ρ is isomorphic to the conjugate $\bar{\rho}$ of ρ. We view v, w, x, y, z as linear functionals on $\mathbf{C}^5$. Therefore the action of G on the span of v, w, x, y, z is by

$$\begin{pmatrix} v \\ w \\ x \\ y \\ z \end{pmatrix} \mapsto {}^t\gamma^{-1} \cdot \begin{pmatrix} v \\ w \\ x \\ y \\ z \end{pmatrix}$$

for all γ in G. The representations ρ and $\bar{\rho}$ are, up to isomorphism, the only

---

[17] If $a = [a_0, \ldots, a_n]$ is a point of n-dimensional projective space and f is a homogeneous polynomial of degree d in $n+1$ variables $x_0, \ldots, x_n$ then the locus

$$\left(a_0 \frac{\partial}{\partial x_0} + \ldots + a_n \frac{\partial}{\partial x_n} \right)^r f = 0$$

in projective space n space is called the r-th polar of a with respect to the hypersurface $f = 0$.. If the left hand side does not vanish identically, the r-th polar is a (possibly nonreduced) hypersurface of degree $d - r$. When $d - r = 2$, the r-th polar is called the **polar quadric** of a with respect to $f = 0$. Clearly, if $r, s \geq 0$, the sth polar with respect to a of the rth polar of a with respect to $f = 0$ is the $(r + s)$th polar of a with respect to $f = 0$.

irreducible representations of G of degree 5. It is useful to introduce the following terminology.

Definition (29.2):   *By a* **standard framing** *of a representation space* $\mathcal{V}$ *of* G *isomorphic to* ρ, *we mean the image of* v, w, x, y, z *under a* G-*isomorphism of the span of* v, w, x, y, z *onto* $\mathcal{V}$.

Denote by B the subgroup of G generated by P and R. The restriction of ρ or $\bar{\rho}$ to B remains irreducible. A standard framing of a representation space $\mathcal{V}$ isomorphic to ρ is therefore a basis u_0, u_1, u_2, u_3, u_4 for $\mathcal{V}$ such that P sends u_j to $exp(2\pi\sqrt{-19}^j/11)u_j$ and R sends u_j to u_{j+1} for $j = 0, 1, 2, 3, 4$, where $j + 1$ is taken modulo 5. Similarly, we can define a standard framing of a representation space isomorphic to $\bar{\rho}$ to be a basis u_0, u_1, u_2, u_3, u_4 such that P sends u_j to $exp(2\pi\sqrt{-12}^j/11)u_j$ and R sends u_j to u_{j+1}.

§30   Invariant Line Bundles

The following result is due to Felix Klein [K2]. Additional commentary on the relevant geometry can be found in Klein-Fricke ([K-F], Bd.II) and in the papers of W.L.Edge [E1] and of the author [A1-A17].

Theorem (30.1):   *The locus* $\mathcal{C}$ *in* $\mathbf{P}^4(\mathbf{C})$ *of all points* $[v, w, x, y, z]$ *for which the matrix*

$$\begin{pmatrix} w & v & 0 & 0 & z \\ v & x & w & 0 & 0 \\ 0 & w & y & x & 0 \\ 0 & 0 & x & z & y \\ z & 0 & 0 & y & v \end{pmatrix}$$

has rank 3 is a smooth curve of degree 20 and genus 26 isomorphic to the modular curve $X(11)$. *Furthermore,* $\mathcal{C}$ *is invariant under the simple group* G, *which is isomorphic to the group* $PSL_2(\mathbf{F}_{11})$ *and which induces on* $\mathcal{C}$ *its full automorphism group.*

Corollary (30.2):   *The restriction of* $\mathcal{O}(1)$ *to* $\mathcal{C}$ *via the inclusion* $j : \mathcal{C} \hookrightarrow \mathbf{P}^4(\mathbf{C})$ *is a locally free sheaf on* $\mathcal{C}$ *associated to a* G-*invariant line bundle* L *on* $\mathcal{C}$.

We will retain the notation λ of §24 for the unique invariant line bundle of degree $(p^2 - 1)/2$ on $X(p)$. In our case, $p = 11$ and the line bundle λ has degree 5. The following result is now an immediate consequence of Theorem 24.1.

Corollary (30.3):   *The canonical bundle* K *on* $X(11)$ *is* λ^{10}. *The invariant line bundle* L *is* λ^4.

Using the remarks and the terminology introduced at the end of §24, we then have the following result.

Corollary (30.4):   *The sections of* L *are modular forms of weight* **4/5** *for* $\Gamma(11)$. *The five homogeneous coordinates of* $\mathbf{P}^4(\mathbf{C})$ *correspond to five linearly independent* modular forms of weight $4/5$ which we denote respectively by $\phi_v, \phi_w, \phi_x, \phi_y, \phi_z$. The linear space spanned by these forms is an irreducible representation for G of degree 5 isomorphic to $\bar{\rho}$ and the forms $\phi_v, \phi_w, \phi_x, \phi_y, \phi_z$ are a standard framing of that space.

Proof: We only need to prove the last statement. The sections of $\mathcal{O}(1)$ on $\mathbf{P}^4(\mathbf{C})$ are just the span of v, w, x, y, z, hence a representation space of type $\bar{\rho}$. Since the restriction of sections to $\mathcal{C}$ is equivariant, the forms $\phi_v, \phi_w, \phi_x, \phi_y, \phi_z$ also span a representation of type $\bar{\rho}$. Since v, w, x, y, z are a standard framing, so are $\phi_v, \phi_w, \phi_x, \phi_y, \phi_z$. When we wish to refer to the system of the five forms $\phi_v, \phi_w, \phi_x, \phi_y, \phi_z$, we will denote it by ϕ.

Corollary (30.5): If $f(v, w, x, y, z)$ is any homogeneous polynomial of degree 5, then

$$f(\phi_v, \phi_w, \phi_x, \phi_y, \phi_z)$$

is a modular form of weight 4.

§31 Construction of modular forms of weight 4/5

In order to compute the forms $\phi_v, \phi_w, \phi_x, \phi_y, \phi_z$, we will choose suitable homogeneous polynomials of degree 5, obtain modular forms of weight 4 that we can recognize and then solve for the original forms of weight $4/5$. First we introduce the homogeneous polynomials of degree 5.

Definition (31.1): We denote by F_v, F_w, F_x, F_y, F_z the homogeneous polynomials of degree 5 defined by

$$\begin{pmatrix} F_v \\ F_w \\ F_x \\ F_y \\ F_z \end{pmatrix} = \left(\begin{pmatrix} v \\ w \\ x \\ y \\ z \end{pmatrix} \#f_3 \right) \#f_7.$$

These forms are all nonzero, since F_v is equal to

$$w^5 + 5 \cdot (-v^3 xz + v^2 w^2 y + v^2 xy^2 + vwx^2 z - 2vwyz^2 - w^3 xy + wy^3 z + x^2 z^3)$$

and the forms F_w, F_x, F_y, F_z are obtained from F_v by successive applications of the cyclic permutation $(vwxyz)$.

Lemma (31.2): The forms F_v, F_w, F_x, F_y, F_z span an irreducible representation of G of degree 5 isomorphic to $\bar{\rho}$. An explicit isomorphism is given by $i \mapsto F_i$, where i runs over v, w, x, y, z.

Proof: This follows at once from the definition and from Lemma 29.1.
Thus, F_v, F_w, F_x, F_y, F_z are a standard framing of the space that they span.

Corollary (31.3):   *The modular forms* $F_v(\phi), F_w(\phi), F_x(\phi), F_y(\phi), F_z(\phi)$ *of weight 4 are a standard framing for an irreducible representation space of type* $\bar{\rho}$ *for G.*

Proof: First we note that the forms $F_v(\phi), \ldots, F_z(\phi)$ are not identically zero. This is easy to see since the explicit form of F_v shows that it contains the term w^5, so it does not vanish on the point $[0, 1, 0, 0, 0]$ of C. Similar remarks apply to the other coordinates. Since the modular forms ϕ map the modular curve $X(11)$ onto C, it follows that the forms $F_v(\phi), \ldots, F_z(\phi)$ are not identically zero. It is easy to see that the forms $F_v(\phi), \ldots, F_z(\phi)$ are cyclically permuted by R. As for their behavior under P, it is clear that they are all eigenforms for P. To determine their eigenvalues, note that the form ϕ_w is mapped to $\zeta^2 \phi_w$ and the quintic F_v contains the term w^5. It follows that $F_v(\phi)$ is mapped to $\zeta^{2 \cdot 5} F_v(\phi) = \zeta^{10} F_v(\phi)$. This determines the eigenvalues, and we are done.

The actual modular forms of weight 4/5 were not available to Klein. Instead, he embedded the modular curve $X(11)$ using a system of 5 functions given on page 404 of [K-F],Bd.II (cf. p.357, equation (2)). Klein and Fricke denote these functions $z_\alpha^{(1)}$ to distinguish them from two other interesting systems of five functions on pp.403,405 which he denotes $z_\alpha^{(3)}$ and $z_\alpha^{(2)}$ respectively. The five functions $z_\alpha^{(1)}$ are actually modular forms for $\Gamma(22)$ which transform with a multiplier of ± 1 under $\Gamma(11)$. Since $\sqrt{\Delta}$ transforms by this same multiplier under the modular group, Klein and Fricke multiply the five functions $z_\alpha^{(1)}$ by $\sqrt{\Delta}$ to obtain modular forms of weight 8 for $\Gamma(11)$ which they denote by $\zeta_\alpha^{(1)}$ on page 405, equation (6). The forms $\zeta_\alpha^{(1)}$ span a representation space isomorphic to $\bar{\rho}$, the explicit transformations being given in (7) on page 405. These forms are not a standard framing, but a simple permutation yields a standard framing which we denote by $\psi_v, \psi_w, \psi_x, \psi_y, \psi_z$. Explicitly, we have

$$z_\alpha^{(1)} = \left(\frac{2\pi}{\omega_2} \right)^2 \cdot \sum{}' (-1)^\xi \cdot \xi \cdot q^{\frac{\xi^2 + 11\xi\eta + 33\eta^2}{22}},$$

where $\sum'$ indicates that the summation runs over all pairs of integers ξ, η such that η is odd and ξ is congruent to $-\alpha$ modulo 11, and

$$\zeta_\alpha^{(1)} = \sqrt{\Delta} \, z_\alpha^{(1)}.$$

Our standard framing is then given by

$$
\begin{aligned}
\psi_v &= \zeta_3^{(1)} \\
\psi_w &= \zeta_9^{(1)} \\
\psi_x &= \zeta_5^{(1)} \\
\psi_y &= \zeta_4^{(1)} \\
\psi_z &= \zeta_1^{(1)}
\end{aligned}
$$

Since these five functions are used to embed $X(11)$ in $\mathbf{P}^4(\mathbf{C})$ and since this embedding induces the line bundle λ^4 whose sections v, w, x, y, z are the modular forms $\phi_v, \phi_w, \phi_x, \phi_y, \phi_z$ of weight 4/5, it follows that there is a meromorphic function γ

on the upper half plane such that

$$\phi_i = \gamma \cdot \psi_i$$

for $i = v, w, x, y, z$. Since the functions ψ_i are known explicitly, we will be done if we can compute γ.

Definition (31.4):   We denote by G_v, G_w, G_x, G_y, G_z the first partial derivatives of f_3 with respect to v, x, w, y, z respectively, or what is the same,

$$\begin{pmatrix} G_v \\ G_w \\ G_x \\ G_y \\ G_z \end{pmatrix} = \begin{pmatrix} v \\ w \\ x \\ y \\ z \end{pmatrix} \# f_3.$$

Thus, G_v is the quadratic form $2vw + z^2$ and G_w, G_x, G_y, G_z are the quadratic forms obtained from G_v by successive applications of the cyclic permutation $(vwxyz)$.

Lemma (31.5):   The quadratic forms G_v, G_w, G_x, G_y, G_z are a standard framing of an irreducible representation of G isomorphic to ρ.

Proof: This follows immediately from the definitions and from Lemma 29.1.

The 5 modular forms which Klein denotes by $z_\alpha^{(3)}$ on page 403 of [K-F], Bd.II are modular forms of weight 2. Explicitly, they are given by

$$z_\alpha^{(3)} = \left(\frac{2\pi}{\omega_2} \right) \cdot \sum{}' \xi \cdot q^{\frac{\xi^2 + 11\xi\eta + 33\eta^2}{11}},$$

where $\sum'$ indicates that the summation runs over all pairs ξ, η of integers such that ξ is congruent to $-\alpha$ modulo 11. They span a representation space of G of dimension 5 isomorphic to ρ and are a standard framing of that space. We will denote them by ω_i, where $i = v, w, x, y, z$. Explicitly, we define

$$(\omega_v, \omega_w, \omega_x, \omega_y, \omega_z) = (z_1^{(3)}, z_3^{(3)}, z_9^{(3)}, z_5^{(3)}, z_4^{(3)}).$$

Since the forms ω_i are modular forms of weight 2, they they are holomorphic sections of the canonical bundle of $X(11)$ or, what is the same, of λ^{10}.

Lemma (31.6):   If $Q(v, w, x, y, z)$ is a quadratic form then $Q(\omega_v, \omega_w, \omega_x, \omega_y, \omega_z)$ is a modular form of weight 4. The modular forms $G_v(\omega), G_w(\omega), G_x(\omega), G_y(\omega), G_z(\omega)$ are a standard framing for an irreducible representation of G of type $\bar{\rho}$.

Proof: The first statement is obvious. Next we observe that the modular forms $G_v(\omega), \dots, G_z(\omega)$ are not identically zero. Indeed, since the forms $G_v, \dots, G_z$ are themselves a standard framing, either they all map the ω to zero or they map them to a standard framing. The former cannot occur because the forms G_v, G_w, G_x, G_y, G_z are easily shown to have no nontrivial common zero, so the five modular forms of weight 4 are a standard framing. One verifies that the eigenvalue of $G_v(\omega)$ under P is ζ^{10} and we are done.

We have now produced two sets of modular forms of weight 4 framing an irreducible representation of G of type ρ. The following lemma is the heart of our computation. If we replace the ϕ_i by the corresponding $z_\alpha^{(1)}$, it gives a remarkable identity among theta functions.

Lemma (31.7):   *Let us denote by* ϕ *the system* $\phi_v, \phi_w, \phi_x, \phi_y, \phi_z$ *of modular forms of weight 4/5 constructed above. Similarly, let us denote by* ω *the system* $\omega_v, \omega_w, \omega_x, \omega_y, \omega_z$ *of modular forms of weight 2 defined above. Then the matrix*

(31.8)
$$\begin{pmatrix} F_v(\phi) & F_w(\phi) & F_x(\phi) & F_y(\phi) & F_z(\phi) \\ G_v(\omega) & G_w(\omega) & G_x(\omega) & G_y(\omega) & G_z(\omega) \end{pmatrix}$$

has rank ≤ 1 *everywhere on the upper half plane. In fact, there is a complex number* c *such that top row is* c *times the bottom row.*

Proof: We can compute the character χ of G acting on the space of all modular forms of weight 4 for $\Gamma(11)$. We do this by identifying such forms with sections of the square of the canonical line bundle of $X(11)$ and using the Woods Hole Fixed Point Formula (SGA 5 III.6.1.2, p.131). If h is an element of G other than the identity then since $H^1(X(11), \Omega^{\otimes 2}) = 0$, the trace of h acting on $H^0(X(11), \Omega^{\otimes 2})$ will be given by

$$\chi(h) = \sum{}' \frac{(dh_x)^2}{1 - (dh_x)},$$

where the summation runs over the fixed points x of h on $X(11)$. For $h = 1$, we have $\chi(h) = 3 \cdot 26 - 3 = 75$, by the Riemann-Roch theorem, since $X(11)$ has genus 26. We therefore find that the character χ is given by

$$\chi(h) = \begin{cases} 75 & \text{if } h = 1 \\ -2 & \text{if } h \text{ has order 11} \\ 3 & \text{if } h \text{ has order 2} \\ 0 & \text{otherwise} \end{cases}$$

We then compute the multiplicity of $\bar{\rho}$ in χ and find that it is equal to 1. Therefore, the two systems of modular forms $F_v(\phi), \ldots, F_z(\phi)$ and $G_v(\omega), \ldots, F_z(\omega)$ must span the same space of modular forms. Since both systems are standard frames, one is a scalar multiple of the other and the lemma is proved.

We can multiply the modular forms ϕ by a suitable constant and assume henceforth that the constant c is 1. We can now compute γ.

Corollary (31.9):   *The function* γ *is given by*

$$\gamma = \sqrt[5]{\frac{G_i(\omega)}{F_i(\psi)}}$$

for any i *in* $\{v, w, x, y, z\}$.

Proof: Since the top row of (31.8) equals the bottom row, we have

$$\gamma^5 \cdot (F_v(\psi), F_w(\psi), F_x(\psi), F_y(\psi), F_z(\psi))$$

$$= \quad (F_v(\phi), F_w(\phi), F_x(\phi), F_y(\phi), F_z(\phi))$$

$$= \quad (G_v(\omega), G_w(\omega), G_x(\omega), G_y(\omega), G_z(\omega)).$$

Therefore for each i in $\{v, w, x, y, z\}$ we have

$$\gamma^5 = \frac{G_i(\omega)}{F_i(\psi)}$$

and the lemma follows at once.

Corollary (31.10): *The system of five modular forms of weight 4/5 for $\Gamma(11)$ is given by*

$$\phi_i = \psi_i \cdot \sqrt[5]{\frac{G_j(\omega)}{F_j(\psi)}}$$

where i, j run over $\{v, w, x, y, z\}$ and where the expression on the right does not depend on the choice of j.

§32   Complements

Let $k = \mathbf{Q}(\cos(2\pi/11))$. Then k is an abelian quintic extension of the field $\mathbf{Q}$ of all rational numbers, namely the real part of the 11-th cyclotomic field. Since k is totally real, it has five infinite places, all of them real. Let ∞ be one of the infinite places. Denote by D the unique (up to isomorphism) quaternion algebra over k which is ramified at all of the infinite places of k except ∞. Let $\mathcal{O}$ be a maximal order of D. One can show that $\mathcal{O}$ is unique up to conjugation by a nonzero element of D. Let $\mathcal{P}$ be the unique prime of k lying over the rational prime 11. Denote by $\Gamma(\mathcal{P})$ the group of all units of $\mathcal{O}$ which have reduced norm 1 and which are congruent to 1 modulo $\mathcal{P}$. Then $\Gamma(\mathcal{P})$ acts on the upper half plane $\mathcal{H}$ without fixed points and with compact quotient. Let $\mathcal{O}^1$ denote the group of all units of $\mathcal{O}$ with reduced norm 1. Then $\mathcal{O}^1/\{\pm 1\} \cdot \Gamma(\mathcal{P})$ is isomorphic to $PSL_2(\mathbf{F}_{11})$. Further, according to [Shu1], p.82, $\mathcal{O}^1$ acts on the upper half plane as a triangle group of type $\{2, 3, 11\}$. It then follows from the Riemann-Hurwitz relation that $\mathcal{H}/\Gamma(\mathcal{P})$ is a curve of genus 26. Since the modular curve of prime level is, according to a theorem of Hecke, uniquely determined by its genus and its automorphism group, it follows that $\mathcal{H}/\Gamma(\mathcal{P})$ is isomorphic to the modular curve $X(11)$. It follows that our investigation of modular forms of weight 4/5 for $\Gamma(11)$ also provide us, at no extra cost, with modular forms of weight 4/5 for the group $\Gamma(\mathcal{P})$.

Appendix III

Invariants and $X(11)$
   by Allan Adler

§33   Introduction

In this note, we show how to describe two different embeddings of the modular curve $X(11)$ into $\mathbf{P}^4$. This is accomplished using the author's description [A8], [A9] of the ring of invariants for $PSL_2(\mathbf{F}_{11})$ acting on $\mathbf{C}^5$. The loci that are obtained are nonreduced but the reduced loci are birational to $X(11)$. What is especially pleasant is the ease with which the equations are obtained given that one knows the ring of invariants. We expect that better results will be obtained once we have obtained the isotypic components of the polynomial ring $\mathbf{C}[v, w, x, y, z]$ under $PSL_2(\mathbf{F}_{11})$ instead of just the invariants. The purpose of this note is merely to illustrate what we regard as a powerful method for studying modular varieties and other varieties with large automorphism groups. We refer to [A9] for notation and conventions regarding the ring of invariants.

§34   The Curve of Degree 20

The modular curve of level 11 can be realized as the singular locus $\mathcal{C}$ of the Hessian of the cubics threefold

$$V^2W + W^2X + X^2Y + Y^2Z + Z^2V = 0.$$

This elegant result is due to Felix Klein [K2],[A2],[A16]. For our purposes, let us assume that $\mathcal{C}_{20}$ is any curve of degree 20 in $\mathbf{P}^4$ invariant under the simple group $G = PSL_2(\mathbf{F}_{11})$ and which is reduced as a scheme.

Lemma (34.1):   *Any G-invariant form of degree 5,7,8 or 10 vanishes identically on $\mathcal{C}_{20}$.*

Proof: If such an invariant does not vanish on $\mathcal{C}_{20}$, it determines a G-invariant positive divisor on $\mathcal{C}_{20}$ having degree 100, 140, 160 or 200. The degree of any positive G-invariant divisor must have the form

$$60a + 220b + 330c + 660d$$

where a,b,c and d are nonnegative integers. But it is straightforward to check that 100, 140, 160 and 200 cannot be written in this way.

Let us now assume that f_6 vanishes identically on C_{20} but that f_3 and f_{11} do not. If we substitute

$$f_5 = f_6 = f_7 = f_8 = f_{10} = 0$$

in the syzygies given in Table V of [A9] (or of [A8]), we find that the syzygy $Syz_{17'}$ of degree 17 becomes $-f_3 f_{14} = 0$, and hence that $f_{14} = 0$ since $f_3 \neq 0$ on C_{20}. From the first syzygy of degree 23, denoted Syz'_{23} in Table V of [A8],[A9], we have that $f_{11} f_{12} = 0$, which implies $f_{12} = 0$ since $f_{11} \neq 0$ on C_{20}. Letting $f_{12} = f_{14} = 0$, we find that the basic relations remaining come from the syzygies Syz'_{21}, Syz'_{24} and Syz_{27}, which become (up to a nonzero scalar factor which does not concern us) respectively

$$-f_3^7 - 2f_3^4 f_9 - f_3 f_9^2 = 0$$

$$f_3^2 \cdot (f_3^6 + 2 \cdot f_3^3 f_9 + f_9^2) = 0$$

$$f_3^9 + 22 \cdot f_3^6 \cdot f_9 + 41 f_3^3 f_9^2 + 20 \cdot f_9^3 = 0$$

From Syz'_{24} and the fact that $f_3 \neq 0$ on C_{20} we have

$$0 = f_3^6 + 2 \cdot f_3^3 \cdot f_9 + f_9^2 = (f_3^3 + f_9)^2.$$

Furthermore, this last equation implies the vanishing of the remaining syzygies. Therefore, if we put

$$f_5 = f_6 = f_7 = f_8 = f_{10} = f_{12} = f_{14}$$

(34.2)

$$f_9 = -f_3^3$$

and if we use the structure of the ring R of invariants, we obtain a homomorphism of R into the ring $\mathbf{C}[f_3, f_{11}]$. Since this ring is a subring of the integral domain R, it follows that the relations (34.2) give rise to a prime ideal in R corresponding to the image of C_{20} under the natural mapping of $\mathbf{P}^4$ onto $Proj(R)$. Since C_{20} is invariant under G, it follows that the relations 34.2 determine the underlying point set of C_{20}, although they do not define a reduced scheme.

§35   The Curve of Degree 50

Let C_{50} be a G-invariant irreducible reduced curve of degree 50 in $\mathbf{P}^4$.

Lemma (35.1):   The invariants f_3, f_5 and f_7 vanish on C_{50}.

Proof: One cannot express 150, 250 or 350 in the form $60a + 110b$, so we are done as in Lemma 1. Now suppose that $\mathcal{C}_{50}$ is the curve obtained from Klein's functions $z_\alpha^{(3)}$ (cf. [K-F], Bd.II,p.403). One checks easily that f_{10} vanishes on $\mathcal{C}_{50}$ while f_6, f_8 and f_9 do not. From the syzygy table, we conclude that on $\mathcal{C}_{50}$ we have

$$f_6 f_{11} = f_8 f_9$$
$$-5 f_6^2 f_8 + 15 f_6 f_{14} - f_8 f_{12} = 0$$
$$5 f_6^2 + 4 f_{12} = 0$$
$$4 f_{14} = f_6 f_8$$
$$864 f_3^3 f_9 + f_8^2 f_{11} + 4 f_9^3 = 0$$

We can therefore define a homomorphism ϕ of R into the ring S, where

$$S = \frac{\mathbf{C}\left[f_6, f_{11}, f_8, f_8^{-1}\right]}{(f_8^5 + 864 f_6^2 f_8^2 + 4 f_6^3 f_{11}^2)},$$

by

$$\phi(f_3) = \phi(f_5) = \phi(f_7) = \phi(f_{10})$$
$$\phi(f_{12}) = -\frac{5}{4} f_6^2$$

(2)

$$\phi(f_{14}) = \frac{1}{4} f_6 f_8$$
$$\phi(f_9) = \frac{f_6 f_{11}}{f_8}$$

The ring S is easily seen to be an integral domain given that f_6, f_8 and f_9 are algebraically independent. Therefore the kernel of ϕ is a prime ideal. It must be the homogeneous prime ideal associated to the image of $\mathcal{C}_{50}$ in $Proj(R)$. The ideal generated by this kernel in $\mathbf{C}[V, W, X, Y, Z]$ must be primary for the prime ideal associated to $\mathcal{C}_{50}$ in $\mathbf{P}^4$. Therefore the relations

$$f_3 = f_5 = f_7 = f_{10} = 0$$
$$f_9 = \frac{f_6 f_{11}}{f_8}$$
$$f_{14} = \frac{1}{4} f_6 f_8$$
$$f_{12} = -\frac{5}{4} f_6^2$$

define a nonreduced scheme whose underlying point set is $\mathcal{C}_{50}$.

Appendix IV

On the Hessian of a Cubic Threefold
by Allan Adler

§36 Introduction

In this paper we will be concerned with some aspects of the geometry of the hessian of a cubic threefold. It is well known that for a generic cubic threefold Λ, the hessian $\mathcal{H}$ has for its singular locus a nonsingular curve $\mathcal{C}$ of degree 20 and genus 26. We feel that the nodal curve of the Hessian of Λ ought to be referred to as the z-**curve** **of** Λ, since its properties generalize those of Klein's z-curve of level 11, which arises in this way from Klein's cubic threefold and is isomorphic to the modular curve $X(11)$. Emphasis on this choice of terminology is also indicated by the fact that it is possible to give a natural generalization, in the context of cubic threefolds, of Klein's A-curve of level 11 as well. This is discussed in Appendix V.

In this paper, we show how to desingularize the hessian. The construction is as follows. The hessian is the locus of points whose polar quadrics with respect to the cubic are cones. The desingularization is the locus $\tilde{\mathcal{H}}$ in $\mathbf{P}^4 \times \mathbf{P}^4$ of all pairs (x, y) such x is a point of the hessian and y is a singular point of the polar quadric of x with respect to the cubic.

For sufficiently generic Λ, there is a natural rank 2 vector bundle on the z-curve of Λ. In particular, one obtains an interesting invariant rank 2 vector bundle on the modular curve $X(11)$.

For a generic point of the hessian, the singular locus of the polar quadric of that point with respect to the cubic is just a point, so that the projection of $\tilde{\mathcal{H}}$ onto its first coordinate maps $\tilde{\mathcal{H}}$ birationally onto $\mathcal{H}$. However, for points of $\mathcal{C}$, the singular locus of the polar quadric is a line quadrisecant to $\mathcal{C}$. The assertion that $\tilde{\mathcal{H}}$ is nonsingular is equivalent to the assertion that there are no self-conjugate triangles formed by the quadrisecants, the notion of conjugacy being explained in §37. The latter assertion is surprisingly delicate and is closely related to the study of the eigenvalues of a certain Hecke operator on the modular curve $X(11)$ of level 11. From this it follows that the Jacobian variety of the z-curve of a sufficiently generic cubic threefold decomposes canonically into the sum of an abelian variety of dimension 10 and one of dimension 16. These abelian varieties are studied further in [A15]. Our methods also allow us to study the question of other self-conjugate polygons formed by the quadrisecants. For example, the number of self-conjugate pentagons is 144.

Apart from the intrinsic interest of an explicit desingularization, recent developments in what is currently called physics make the Hessian an attractive example on which to do some experiments in mirror symmetry. It has been known for a few years that one can study quintic threefolds from this point of view, but these quintic threefolds have been assumed to be nonsingular. The hessian, however, is a singular quintic. As such, it is a degeneration of nonsingular quintics and might have a useful interpretation in terms of the boundary of the moduli space of Calabi-Yau quintic hypersurfaces. It is therefore tempting to see whether the results for nonsingular quintics can be extended to the desingularization of the hessian. For this purpose one needs to have one's hands on a desingularization. From the description of $\tilde{\mathcal{H}}$, it is easy to see that $\tilde{\mathcal{H}}$ is a Calabi-Yau manifold. We are therefore perfectly situated to do this kind of experimental "physics". The result, mentioned in the last paragraph, about the decomposition of the Jacobian of the nodal curve of the Hessian then has implications for the periods of the Calabi-Yau manifold $\tilde{\mathcal{H}}$.

We remark that although this explicit desingularization is quite interesting in its own right, it is also a lot of work to construct it. Therefore, it is pleasant to know that in recent years considerable progress (cf. [FH], [FHP], [Has1-3], [Pa-St], [Sa], [St1-2]) has been made on the problem of computing birational invariants of a desingularization of a given variety without having to construct such a desingularization. It is our understanding that despite that progress, the example considered in the present article is still beyond the methods of L^2-cohomology and intersection cohomology. In view of the resources presented here for studying these examples, an examination of the desingularization of the hessian of a sufficiently generic cubic threefold from the standpoint of L^2-cohomology and intersection cohomology could prove to be both feasible and fruitful.

In §40, we discuss the role of the desingularization of the Hessian in the study of a compact Shimura variety of dimension 3, as well a role for the more general methods of L^2-cohomology mentioned in the preceding paragraph.

In §37 and §42, we have touched briefly on an old article [G-N] of Gordan and Noether, proving only as much as we needed for the present article. However, their beautiful and ingenious article deserves a more thorough exposition and we plan to give it one.

The author's results on the desingularization of the hessian of a sufficiently generic cubic threefold were obtained in 1974-76 while the author was visiting the Institute for Advanced Study in Princeton and described in the author's unpublished manuscript [A2]. While the author was visiting IHES during 1994-95, he learned about the notion of mirror symmetry from Mischa Gromov and realized that his old work on desingularizing the hessian was related to the phenomenon. Although the desingularization is of independent interest and the author was planning to publish it anyway, the relation to mirror symmetry makes the work of interest to a wider community of mathematicians.

The author is grateful to André Weil for introducing him to the cubic threefold

$$v^2w + w^2x + x^2y + y^2z + z^2v = 0$$

which has so greatly influenced the author's work for the last 20 years. The author is grateful to Mischa Gromov for introducing him to the concept of mirror symmetry in 1994. Finally, the author thanks Masake Kashiwara for mentioning the paper [G-N] of Gordan and Noether to him in 1976-78.

§37 Singular Locus vs. Rank ≤ 3 Locus

Denote by f a cubic form in the 5 variables[18] x_0, x_1, x_2, x_3, x_4 with coefficients in the field $\mathbf{C}$ of complex numbers. We will also write

$$x = (x_0, x_1, x_2, x_3, x_4).$$

We denote by $M(f)$ the matrix of second partial derivatives of f. If we wish to emphasize the dependence of $M(f)$ on x, we will write $M_x(f)$. If $a = (a_0, a_1, a_2, a_3, a_4)$ is a point of $\mathbf{C}^5$, we will denote by $M_a(f)$ the matrix obtained by replacing x_i by a_i for $0 \leq i \leq 4$. In most cases, we will be concerned with properties of the matrix $M(f)$, such as its rank, which are homogeneous in x. Accordingly, if we have a point $[a] = [a_0, a_1, a_2, a_3, a_4]$ of $\mathbf{P}^4$, we will freely write $M_a(f)$ even though the homogeneous coordinates a_0, a_1, a_2, a_3, a_4 of $[a]$ are only well defined up to a scalar multiple. I will also tend to abuse notation by failing to distinguish between row vectors and column vectors. Thus, if $b = (b_0, b_1, b_2, b_3, b_4)$ is a point of $\mathbf{C}^5$, or even a point of $\mathbf{P}^4$, I will write $M_a(f)b$ and $bM_a(f)$ for the product of the matrix $M_a(f)$ and the vector $(b_0, b_1, b_2, b_3, b_4)$. It doesn't really matter which side the vector is placed on since the matrix $M_a(f)$ is symmetric. Furthermore, we can regard $M_x(f)$ as a 3-tensor by the rule

$$(a, b, c) \mapsto aM_b(f)c.$$

This 3-tensor is symmetric and for $a = b = c = x$ is a constant multiple of the original cubic form f.

The determinant of $M(f)$ is a quintic form in x called the **Hessian** of f and will be denoted $\mathcal{H}(f)$. If we denote by Λ the hypersurface $f = 0$ in $\mathbf{P}^4$, the locus $\mathcal{H}(f) = 0$ will be called the Hessian of Λ and denoted $\mathcal{H}(\Lambda)$. If the Hessian of f is not identically zero, the Hessian of Λ is a hypersurface. We remark that the Hessian of Λ really does depend only on the point set Λ and not on the choice of the form f. Indeed, the only way in which f can fail to be determined by Λ up to a constant factor is for Λ to consist of two hyperplanes of which one is counted twice. In that case, after a linear change of coordinates, Λ is defined set-theoretically by either of the two forms $x_0^2 x_1$ and $x_0 x_1^2$. But for both of these forms, the Hessian vanishes identically and therefore depends only on the point set Λ.

Throughout this paper, we will use the notation f_0 to denote Klein's cubic form ([K2] §4, p.147; [K-F] Vol.II, p.410)

$$(37.1) \qquad\qquad v^2w + w^2x + x^2y + y^2z + z^2v$$

and Λ_0 to denote the cubic hypersurface $f_0 = 0$. We will occasionally wish to refer

---

[18] We will often use v, w, x, y, z instead of x_0, x_1, x_2, x_3, x_4.

to the Fermat hypersurface $f_1 = 0$, where

$$(37.2) \qquad\qquad f_1 = v^3 + w^3 + x^3 + y^3 + z^3.$$

In all of these cases, the cubic hypersurface is nonsingular. We will consider only nonsingular cubics for the rest of this paper. Henceforth f will refer to an arbitrary nonsingular cubic form, subject to possible further restrictions depending on the context, and Λ will denote the cubic hypersurface $f = 0$.

The following result will be proved in §42. The proof relies on methods of Gordan and Noether [G-N].

Theorem (37.3):   *The hessian of a nonsingular form of degree ≥ 2 doesn't vanish identically.*

In general, we will denote by $C(f)$ or $C(\Lambda)$ or simply C the singular locus of $\mathcal{H}(\Lambda)$. The singular locus C is defined by the first partial derivatives of $\mathcal{H}(f)$. Each of these first partials is a linear combination of 4×4 minors of $M(f)$. Therefore if all of the 4×4 minors of $M_a(f)$ vanish for a point $[a]$ of $\mathbf{P}^4$ then $[a]$ lies in the singular locus C of $\mathcal{H}(\Lambda)$. We will refer to the locus in $\mathbf{P}^4$ defined by the 4×4 minors as the **rank ≤ 3 locus of the Hessian**. Thus, the singular locus contains the rank ≤ 3 locus. We will prove that the opposite inclusion holds when the cubic Λ is nonsingular. However, first we introduce a construction which will be useful both in that proof and in the rest of the paper.

Denote by $\widehat{\mathcal{H}(\Lambda)}$ the locus in $\mathbf{P}^4 \times \mathbf{P}^4$ consisting of all $([a], [b])$ for which $M_a(f)b = 0$. Denote by π_1 and π_2 the projections of $\widehat{\mathcal{H}(\Lambda)}$ onto the first and second factors of $\mathbf{P}^4 \times \mathbf{P}^4$ respectively.

Lemma (37.4):   *Let f be a nonsingular quinary cubic form. Then the singular locus of $\mathcal{H}(\Lambda)$ is equal to the rank ≤ 3 locus.*

Proof: Let U be the complement of C in $\mathcal{H}(\Lambda)$. In other words, U is the Zariski open subset of $\mathcal{H}(\Lambda)$ where $M(f)$ has rank equal to 4. Then π_1 is an isomorphism over U. Therefore, it is enough to show that if (p, q) is a point of $\widehat{\mathcal{H}(\Lambda)}$ such that p lies in U, then $\widehat{\mathcal{H}(\Lambda)}$ is nonsingular at (p, q).

Denote by $\widehat{\widehat{\mathcal{H}(\Lambda)}}$ the locus of all points (x, y) of $\mathbf{C}^5 \times \mathbf{C}^5$ which satisfy the system of 5 bilinear equations

$$M_x(f)y = 0.$$

If $[x]$ is a point of $\mathbf{P}^4$ or, what is the same, a line through the origin of $\mathbf{C}^5$, then the tangent space to $\mathbf{P}^4$ at $[x]$ may be identified with the space $Hom([x], \mathbf{C}^5/[x])$. Therefore, the tangent space to $\mathbf{P}^4 \times \mathbf{P}^4$ at the point $([p], [q])$ may be identified with the space

$$Hom([p], \mathbf{C}^5/[p]) \oplus Hom([q], \mathbf{C}^5/[q]).$$

We can compute the tangent space to a point $([p], [q])$ of $\widehat{\mathcal{H}(\Lambda)}$ by the usual Jacobian criterion used for subvarieties of projective space. Computing the Jacobian matrix J of the 5 bilinear equations defining $\widehat{\mathcal{H}(\Lambda)}$, we find that J is a 5×10 matrix which is the concatenation of the two matrices $M_q(f)$ and $M_p(f)$, i.e.:

$$J = (M_q \ M_p).$$

The space T is then the space of all elements (x, y) of $\mathbf{C}^5 \oplus \mathbf{C}^5$ such that

$$M_q(f)x = M_p(f)y = 0.$$

In particular, $[p] \oplus [q]$ is a subspace of T. Accordingly, $([p], [q])$ is a simple point of $\widehat{\mathcal{H}(\Lambda)}$ if and only if (p, q) is a simple point of $\widehat{\mathcal{H}(\Lambda)}$.

Now, (p, q) is a singular point of $\widehat{\mathcal{H}(\Lambda)}$ if and only if the rank of J is less than 5 or, what is the same, if and only if there is an element $[r]$ of $\mathbf{P}^4$ such that

(37.5) $M_p(f)r = M_q(f)r = 0.$

Since we also have $M_p(f)q = 0$ and since we have assumed that $M_p(f)$ has rank 4, it follows that $[r] = [q]$, hence the equation $M_q(f)r = 0$ implies

$$M_q(f)q = 0.$$

But $M_q(f)q$ is, up to a trivial factor of 2, the vector of first partial derivatives of f at q. Therefore, if $([p], [q])$ is a singular point of $\widehat{\mathcal{H}(\Lambda)}$ then $[q]$ is a singular point of Λ. Since we assume that Λ is nonsingular, we are done.

We note that if the cubic is singular at a point $[p]$ and the Hessian $\mathcal{H}(f)$ does not vanish identically, then the point $([p], [p])$ will be singular point of $\widehat{\mathcal{H}(\Lambda)}$. However, we will not need this fact.

A close examination of the proof of the preceding lemma shows that we have the following criterion the locus $\widehat{\mathcal{H}(\Lambda)}$ to be singular. To formulate it, we introduce a convenient notation. For $[x]$ in $\mathbf{P}^4$, denote by $\iota([x])$ the locus of all points $[y]$ of $\mathbf{P}^4$ such that

$$M_x(f)y = 0.$$

Thus, $\iota([x])$ is empty if $[x]$ does not lie in $\mathcal{H}(\Lambda)$. It is a singleton if $[x]$ lies in $\mathcal{H}(\Lambda)$ but not in the rank ≤ 3 locus of $\mathcal{H}(\Lambda)$. It is a line if $[x]$ lies in the rank ≤ 3 locus but not in the rank ≤ 2 locus, and so on.

Lemma (37.6):   *Assume that the cubic Λ is nonsingular. The locus $\widehat{\mathcal{H}(\Lambda)}$ in $\mathbf{P}^4 \times \mathbf{P}^4$ is singular if and only if there exist three points $[p], [q], [r]$ of $\mathbf{P}^4$ such that*

$$[p] \in \iota([q])$$
$$[q] \in \iota([r])$$
$$[r] \in \iota([p]).$$

We will be particularly interested in the case where the rank ≤ 2 locus of $\mathcal{H}(\Lambda)$ is empty. In that case we have the following statement which has a much more distinctly geometrical flavor.

Lemma (37.7):   *Suppose that the cubic Λ is nonsingular and that the rank ≤ 2 locus of $\mathcal{H}(\Lambda)$ is empty. Then $\widehat{\mathcal{H}(\Lambda)}$ is singular if and only if there exist 3 points*

$[p], [q], [r]$ of $\mathbf{P}^4$ which are the vertices of a triangle whose vertices are $\iota([p])$, $\iota([q])$, $\iota([r])$.

§38   Generalization of Klein's z-curve

If f is a nonzero cubic form in 5 variables, denote by $P(f)$ the following assertion about f:

> The Hessian $\mathcal{H}(f)$ of f does not vanish identically and the singular locus of the hypersurface $\mathcal{H}(f)$ of $\mathbf{P}^4$ is a smooth irreducible curve of degree 20 and genus 26.

Since the property $P(f)$ depends only on the cubic hypersurface Λ, we will also write $P(\Lambda)$. The set of all possible Λ is a projective space $\mathbf{P}(S_3(\mathbf{C}^5))$ of dimension 34 which I will denote $\mathcal{K}$. In this section, we will sketch the proof of the following well known result.

Theorem (38.1):   Denote by U the set of all Λ in $\mathcal{K}$ for which $P(\Lambda)$ holds. Then U is a Zariski open subset of $\mathcal{K}$ containing Klein's cubic threefold Λ_0.

Proof: Denote by $P_i(\Lambda)$, for $0 \leq i \leq 5$, the following properties.
  $P_0(\Lambda)$:   The Hessian $\mathcal{H}(\Lambda)$ of Λ is a hypersurface.
  $P_1(\Lambda)$: $P_0(\Lambda)$ and the singular locus of $\mathcal{H}(\Lambda)$ is a curve.
  $P_2(\Lambda)$: $P_1(\Lambda)$ and the singular locus of $\mathcal{H}(\Lambda)$ is irreducible.
  $P_3(\Lambda)$: $P_2(\Lambda)$ and the singular locus of $\mathcal{H}(\Lambda)$ has degree 20.
  $P_4(\Lambda)$: $P_3(\Lambda)$ and the singular locus of $\mathcal{H}(\Lambda)$ is nonsingular.
  $P_5(\Lambda)$: $P_4(\Lambda)$ and the singular locus of $\mathcal{H}(\Lambda)$ has genus 26.
    Thus, the property P_5 is the same as P. Since the identical vanishing of the Hessian is obviously a closed condition on Λ, condition $P_0(\Lambda)$ is clearly an open condition. We now proceed by a sequence of lemmas.

Lemma (38.2):   Condition P_1 is open and $P_1(\Lambda_0)$ holds.

Proof: Denote by $\mathcal{Q}$ the projective space associated to the space of symmetric 5×5 matrices with entries in the field $\mathbf{C}$ of complex numbers. If B is a nonzero 5×5 symmetric matrix then the corresponding element of $\mathcal{Q}$ will be denoted $[B]$. For $k \leq 5$, denote by $\mathcal{Q}_k$ the locus of all $[B]$ in $\mathcal{Q}$ for which the rank of B is $\leq k$ and by $\mathcal{Q}'_k$ the Zariski open subset $\mathcal{Q}_k - \mathcal{Q}_{k-1}$ of $\mathcal{Q}_k$. One can show that the subsets $\mathcal{Q}'_k$ are precisely the orbits of $PSL_5(\mathbf{C})$ acting on $\mathcal{Q}$ and from this one can determine the dimensions of the loci $\mathcal{Q}_k$. It is also easy to see from the $PSL_5(\mathbf{C})$ action and from the fact that the first partials of $(k+1) \times (k+1)$ minors are linear combinations of $k \times k$ minors that $\mathcal{Q}_{k-1}$ is the singular locus of $\mathcal{Q}_k$ for $k \leq 4$. One can explicitly resolve the singularities of $\mathcal{Q}_k$ as follows: define the rational mapping

$$\phi_k : \mathcal{Q}_k \to Gr(5-k, 5),$$

where $Gr(5-k, 5)$ denotes the Grassmannian of $5-k$-planes in $\mathbf{C}^5$, by the rule

$$\phi_k(B) = ker(B),$$

where $ker(B)$ denotes the null space of the matrix B. The rational mapping Φ_k is well defined precisely on the Zariski open subset $\mathcal{Q}'_k$ of $\mathcal{Q}_k$. If we pass to the Zariski closure of the graph of ϕ_k, we obtain the subset $\widehat{\mathcal{Q}}_k$ of $\mathcal{Q}_k \times Gr(5-k,5)$ defined as follows:

$$\widehat{\mathcal{Q}}_k = \{([B],L) \mid L \subseteq ker(B)\}.$$

On the other hand, by considering the projection of $\widehat{\mathcal{Q}}_k$ onto $Gr(5-k,5)$, we see that $\widehat{\mathcal{Q}}_k$ is isomorphic to the projectivized vector bundle

$$\mathbf{P}(S_2(\mathbf{C}^5/E)),$$

where $\mathbf{C}^5$ denotes the trivial 5 dimensional vector bundle, E denotes the tautological $5-k$-plane bundle, which is a subbundle of $\mathbf{C}^5$. In particular, $\widehat{\mathcal{Q}}_k$ is nonsingular as predicted. Since projection of $\widehat{\mathcal{Q}}_k$ onto $\mathcal{Q}_k$ is an isomorphism over $\mathcal{Q}'_k$, in particular $\mathcal{Q}_k$ and $\widehat{\mathcal{Q}}_k$ have the same dimension. Using the description as a projectivized vector bundle, we can easily compute that dimension to be

$$k(5-k) + \frac{k(k+1)}{2} - 1 = \frac{11k - k^2 - 2}{2},$$

which can be expressed more simply by saying that the codimension of $\mathcal{Q}_k$ in $\mathcal{Q}$ is

$$\frac{(5-k)(6-k)}{2}.$$

In particular, the dimension of $\mathcal{Q}_3$ is $14-3 = 11$. On the other hand, we can regard the matrix $M_x(f)$ as a linear mapping from $\mathbf{C}^5$ to $S_2(\mathbf{C}^5)$. If that linear mapping is not injective then there is a point $[p]$ of $\mathbf{P}^4$ such that

$$M_p(f) = 0,$$

which would imply that

$$M_p(f)q = 0$$

for all q in $\mathbf{C}^5$. This in turn would imply that

$$M_q(f)p = 0$$

for all q in $\mathbf{C}^5$, which would imply that the Hessian of f vanishes identically. Therefore, whenever condition P_0 holds, the mapping

$$x \mapsto M_x(f)$$

is an injective linear mapping from $\mathbf{C}^5$ into $S_2(\mathbf{C}^5)$. It therefore defines a linear embedding of $\mathbf{P}^4$ into $\mathcal{Q}$. Denote by H the image of that linear embedding. Then H is a 4 plane in the 14 dimensional projective space $\mathcal{Q}$ and that 4 plane necessarily meets the 11 dimension locus $\mathcal{Q}_3$ in a locus all of whose components are of dimension $\geq 11 + 4 - 14 = 1$. The locus of all Λ such that the intersection has at least one component of dimension at least 2 is therefore closed in $\mathcal{K}$. We will be done if we can show that it is not all of $\mathcal{K}$. To see that, consider Klein's cubic Λ_0 given by

$$v^2w + w^2x + x^2y + y^2z + z^2v = 0.$$

Denote by $\mathcal{C}_0$ its rank ≤ 3 locus. To prove that every component of $\mathcal{C}_0$ is of dimension 1, we will show that the intersection of $\mathcal{C}_0$ with the hyperplane $v = 0$ is a finite set. We will make good use of the invariance of Λ_0 under the cyclic permutation $(vwxyz)$. Thus, we want to show that up to scalar multiples, there are only finitely many matrices

$$\begin{pmatrix} w & v & 0 & 0 & z \\ v & x & w & 0 & 0 \\ 0 & w & y & x & 0 \\ 0 & 0 & x & z & y \\ z & 0 & 0 & y & v \end{pmatrix}$$

with $v = 0$ and rank ≤ 3. Put $v = 0$. The minor obtained by deleting the last row and the second column is then

$$\begin{vmatrix} w & 0 & 0 & z \\ v & w & 0 & 0 \\ 0 & y & x & 0 \\ 0 & x & z & y \end{vmatrix} = w^2xy.$$

So if $v = 0$, then w, x or y is also zero. It follows from the cyclic symmetry of the locus $\mathcal{C}_0$ that if one coordinate vanishes then at least two vanish. After applying a suitable power of that cyclic symmetry, we can assume that when $v = 0$ the other vanishing coordinate is either w or x. In the first case, we put $v = w = 0$ and consider the 4×4 minor obtained by deleting the fourth row and column and obtain

$$\begin{vmatrix} w & v & 0 & z \\ v & x & w & 0 \\ 0 & w & y & 0 \\ z & 0 & 0 & v \end{vmatrix} = -xyz^2,$$

so at least 3 coordinates vanish in this case. In the second case, we put $v = x = 0$ and find that the 4×4 minor in the upper left hand corner is

$$\begin{vmatrix} w & v & 0 & 0 \\ v & x & w & 0 \\ 0 & w & y & x \\ 0 & 0 & x & z \end{vmatrix} = -w^3z,$$

so in either case, at least 3 coordinates vanish as soon as one coordinate does. Applying a suitable power of the cyclic symmetry, we can assume that when $v = 0$, the other two vanishing coordinates are either w, x or w, y. In the first case, we put $v = w = x = 0$ and we find that the 4×4 minor obtained by deleting the second row and column is

$$\begin{vmatrix} w & 0 & 0 & z \\ 0 & y & x & 0 \\ 0 & x & z & y \\ z & 0 & y & v \end{vmatrix} = -yz^3,$$

which forces yet another coordinate to vanish. In the second case, we put $v = w = y = 0$ and find that the 4×4 minor obtained by deleting the third row and column is

$$\begin{vmatrix} w & v & 0 & z \\ v & x & 0 & 0 \\ 0 & 0 & z & y \\ z & 0 & y & v \end{vmatrix} = -xz^3,$$

so in any case a fourth coordinate must vanish. Therefore, as soon as one coordinate vanishes, four of them do. But there are only 4 points in $\mathbf{P}^4$ with 4 vanishing coordinates, including the first. This proves $P_1(\Lambda_0)$ and $P_1(\Lambda)$.

We note that for $P_1(\Lambda)$ to hold for a cubic Λ, it is not sufficient that Λ be nonsingular. Indeed, the diagonal cubic $f_1 = 0$ of (37.2) doesn't satisfy P_1.

Lemma (38.3): P_2 *is an open condition and* $P_2(\Lambda_0)$ *holds.*

Proof: Irreducibility is always an open condition. As for the assertion $P_2(\Lambda_0)$, we note that since the hyperplane $v = 0$ meets C_0 in only 4 points, C_0 can have at most 4 connected components. Actually, each connected component must be irreducible because each irreducible component must meet $v = 0$ and can only do so at one of these four points, but, as we will show, each of these 4 points is a simple point of C_0 and can have at most one irreducible component through it. Using the cyclic symmetry of the locus C_0, it is enough to show that the point $p = [1, 0, 0, 0, 0]$ is a simple point. Now, a monomial of degree 3 in v, w, x, y, z will vanish at p unless that monomial is v^3. Therefore, if N is a 4×4 minor of $M(f_0)$, all of the first partials of $det(N)$ will vanish at p unless N contains the 3×3 matrix obtained by deleting the third and fourth rows and columns of $M(f_0)$. It follows that the tangent space to C_0 at p is the intersection of the tangent spaces at p of

$$K_1 : \quad \begin{vmatrix} w & v & 0 & z \\ v & x & w & 0 \\ 0 & w & y & 0 \\ z & 0 & 0 & v \end{vmatrix} = 0$$

$$K_2 : \quad \begin{vmatrix} w & v & 0 & z \\ v & x & w & 0 \\ 0 & 0 & x & y \\ z & 0 & 0 & v \end{vmatrix} = 0$$

$$K_3 : \quad \begin{vmatrix} w & v & 0 & z \\ v & x & 0 & 0 \\ 0 & 0 & z & y \\ z & 0 & y & v \end{vmatrix} = 0.$$

One verifies that the tangent space to K_1 at p is given by $y = 0$, the tangent space to K_2 at p is given by $x = 0$ and the tangent space to K_3 at p is given by $z = 0$. Therefore the tangent space to C_0 at p is the line $x = y = z = 0$. In particular, the tangent space is of the same dimension as the locus C_0, which is 1. This proves that p is a simple point of C_0 and in particular that C_0 has at most 4 irreducible components.[19] These 4 components are equivalent under cyclic permutations of the coordinates, since C_0 is invariant under such permutations and since the 4 points in question are equivalent under them. However, the cyclic permutations form a cyclic group of order 5 and if this group acts on the ≤ 4 components, it must leave each of them invariant. Since they are equivalent under the action of this group, it follows that there is exactly one component. This completes the proof of the lemma.

Lemma (38.4):   *The condition P_3 is open and $P_3(\Lambda_0)$ holds.*

Proof: We can identify C_0 with the intersection of the linear variety H with Q_3 in Q, as in Lemma 2.2. Since the intersection has dimension $1 = dim(H) + dim(Q_3) - dim(Q)$, it follows that the degree of the intersection, with mulplicities counted, is the product of the degrees of H and of Q_3. The condition that the intersection be transverse, with no multiple components, is an open condition and for such Λ satisfying P_2, the intersection will have degree equal to the degree of Q_3. So to prove that P_3 is an open condition, it suffices to prove that Q_3 has degree 20. This is well known and follows from formulas due to Giambelli [Gi1] (cf.[Har-Tu]). As for showing that P_3 holds for Λ_0, we note first of all that its degree is certainly ≤ 20. The only way it can be < 20 is for C_0 to have multiple components, which it doesn't since it is irreducible and $p = [1, 0, 0, 0, 0]$ is a simple point. Therefore, P_3 holds for Λ_0.

Lemma (38.5):   *The condition P_4 is an open condition and $P_4(\Lambda_0)$ holds.*

Proof: Nonsingularity is always an open condition. As for Λ_0, suppose that C_0 is singular. Since Λ_0 is an invariant hypersurface for the action of $PSL_2(\mathbf{F}_{11})$ on $\mathbf{P}^4$, the singular locus of C_0 must be invariant under that group as well. Since the group has no orbit in $\mathbf{P}^4$ with < 60 points. So if there is one singular point, there are at least 60. Since the space of cubic forms in 5 variables has dimension 35, given any 34 points of $\mathbf{P}^4$, there is a nonzero cubic form which vanishes at all 34 points. In particular, if C_0, we can find a nonzero cubic form F which vanishes at 34 singular points of C_0. Since each of these 34 singular points contributes at least 2 intersections, the intersection of the cubic hypersurface $F = 0$ with C_0 would have at least 68 intersections, counting multiplicity. Since this is $> 3 \times 20 = 60$, that would imply that C_0 is contained in the cubic $F = 0$. So we will be done if we can prove that C_0 doesn't lie in any cubic hypersurface. The space of cubics which do vanish on C_0 is invariant under the group $PSL_2(\mathbf{F}_{11})$. It must therefore be a direct sum of irreducible subspaces. Since the representation of $PSL_2(\mathbf{F}_{11})$ on $S_3(\mathbf{C}^5)$ decomposes as the sum of irreducible representations of degrees 1, 10, 12, 12, it would follow that all of the cubics in one of these irreducible subspaces vanishes on C_0. The cubic in the trivial representation is none other than Λ_0, which we already know does not contain C_0. Each of the remaining representations contains a unique, up to constant multiple, cubic form which is an eigenvector with eigenvalue ζ for the substitution given by the diagonal matrix P with diagonal entries $\zeta, \zeta^9, \zeta^4, \zeta^3, \zeta^5$, where ζ is a primitive 11th root of unity. Therefore, if there is a cubic form vanishing

---

[19] The argument which follows for proving the irreducibility of C_0 was pointed out to the author by Marc Levine. It is much simpler than the argument originally used by the author.

on C_0, there must be one which does so and which is a ζ-eigenfunction of P. The most general cubic form which is a ζ-eigenfunction of P is

(38.6)
$$\alpha w^2 z + \beta xyz + \gamma x^3.$$

Assume such a cubic vanishes on C_0. Then it must vanish at the point $q = [0,0,1,0,0]$ of C_0, so $\gamma = 0$. The cubic form is therefore the product of the linear form z and the quadratic form

(38.7)
$$\alpha w^2 + \beta xy.$$

Since the 5 points obtained from q by cyclic permutation of the coordinates are on C_0 and since these 5 points span all of $\mathbf{P}^4$, C_0 does not lie in any hyperplane, so in particular z cannot vanish on C_0. Since C_0 is irreducible, it follows that the quadratic form $\alpha w^2 + \beta xy$ vanishes on C_0. It must therefore vanish at the point $[0,1,0,0,0]$, so $\alpha = 0$. If $\beta \neq 0$, then xy must vanish on C_0, which implies that either x does or y does. But neither does since, as we noted, C_0 does not lie in any hyperplane. This concludes the proof of the lemma.

Lemma (38.8): *The condition P_4 is open and $P_4(\Lambda_0)$ holds.*

Proof:[20] In the proof of the preceding lemma, we showed that C_0 doesn't lie in any cubic hypersurface. Furthermore, not lying in any cubic hypersurface is an open condition, so the cubics Λ which satisfy P_3 and for which C doesn't lie in any cubic form a nonempty Zariski open set of cubics. Let Λ be such a cubic. Denote by D a divisor of degree 60 cut out on C by the cubic Λ. Since the space of cubic forms in 5 variables has dimesion 35, if we are given any positive divisor E of degree 34 on C, we can find a cubic hypersurface Φ whose intersection with C is a positive divisor D' which is $\geq E$. The divisor $D' - E$ is a positive divisor of degree 26 which we will denote E'. Since D and D' are both cut out by cubics, they are linearly equivalent. Therefore, if we take a base point p of C and use angle brackets to denote the element of the Jacobian variety $Jac(C)$ determined by a divisor of degree 0, we have

$$\langle E - 34p \rangle = \langle D' - E' - 34p \rangle = \langle D - E' - 34p \rangle = \langle D - 60p \rangle - \langle E' - 26p \rangle.$$

This proves that the subset of $Jac(C)$ obtained from positive divisors of degree 34 under the mapping

$$\Delta \mapsto \langle \Delta - deg(\Delta)p \rangle$$

really depends only on 26 parameters. Therefore the Jacobian variety of C has dimension ≤ 26 and the genus of C is ≤ 26. Since the genus does not increase under specialization, it follows that the cubics Λ for which P_4 holds and C has genus exactly 26 will be Zariski open. It remains to show that C_0 has genus ≥ 26. Let g_0 be the genus of C_0 and suppose that $g_0 \leq 26$. Denote by $\mathcal{P}$ the quotient of C_0 by the action of $PSL_2(\mathbf{F}_{11})$ and let g be the genus of $\mathcal{P}$. Let $p_1, \ldots, p_n$ be the branch points on $\mathcal{P}$ of the branched covering $C_0 \to \mathcal{P}$ and let $e_1, \ldots, e_n$ be their orders of ramification. Then p_i has $660/e_i$ preimages, since 660 is the order of the group

---

[20] This argument based on Clifford's Theorem [Cl1] is due to W.L.Edge [E1].

$PSL_2(\mathbf{F}_{11})$. By the Riemann-Hurwitz formula, we therefore have

$$2 - 2g_0 = 660(2 - 2g) - \sum_{i=1}^{n} 660(1 - \frac{1}{e_i}).$$

Dividing both sides by -1320, we get

$$\frac{25}{660} \geq \frac{g_0 - 1}{660} = g - 1 + \frac{1}{2}\sum_{i=1}^{n}(1 - \frac{1}{e_i}).$$

Since the symmetry P fixes the point $p = [1, 0, 0, 0, 0]$ of C_0, we have $n \geq 1$ and one of the e_i, say, e_1 is 11. If $g \geq 1$, we have

$$g - 1 + \frac{1}{2}\sum_{i=1}^{n}(1 - \frac{1}{e_i}) \geq \frac{n}{4} \geq \frac{1}{4} > \frac{25}{660},$$

which is impossible. Therefore $g = 0$. On the other hand, the genus g_0 of C_0 cannot be 0 since $SL_2(\mathbf{F}_{11})$ has no complex projective representation of dimension 2. We also can't have $g_0 = 1$ since the group $SL_2(\mathbf{F}_{11})$ has no nontrivial representation into the group of affine transformations of the affine line over $\mathbf{C}$. We must also have $n \geq 3$ since otherwise the fundamental group of $\mathcal{P} - \{p_1, \ldots, p_n\}$ would be abelian and the branched covering $C_0 \to \mathcal{P}$ could not exist. We have

$$\frac{25}{660} \geq \frac{g_0 - 1}{660} = -1 + \frac{1}{2}\sum_{i=1}^{n}(1 - \frac{1}{e_i}) = -\frac{6}{11} + \frac{1}{2}\sum_{i=2}^{n}(1 - \frac{1}{e_i}) \geq -\frac{6}{11} + \frac{n-1}{4}$$

and if $n > 3$ we have

$$\frac{25}{660} \geq -\frac{6}{11} + \frac{3}{4} = \frac{9}{44},$$

which is a contradiction. Therefore $n = 3$. Since $e_1 = 1$, we have

$$\frac{25}{330} = -\frac{12}{11} + 1 - \frac{1}{e_2} + 1 - \frac{1}{e_3} - \frac{1}{11} - \frac{1}{e_2} + 1 - \frac{1}{e_3}$$

or, what is the same,

$$\frac{1}{e_2} + \frac{1}{e_3} = \frac{5}{6},$$

which implies, since we can assume $e_2 \leq e_3$, that $e_2 = 2$ and $e_3 = 3$. Now that we know e_1, e_2, e_3, we can return to the Riemann-Hurwitz formula and conclude that

$$2 - 2g_0 = 1320 - 600 - 330 - 440 = -50,$$

whence $g_0 = 26$ and the lemma is proved. This also completes the proof of Theorem 38.1.

Theorem (Klein) (38.9):   *The singular locus C_0 of the nonsingular cubic threefold Λ_0 given by*

$$v^2w + w^2x + x^2y + y^2z + z^2v = 0$$

is isomorphic to the modular curve $X(11)$ of level 11.

Proof: We have established that C_0 is a smooth curve of genus 26 which admits the group $PSL_2(\mathbf{F}_{11})$ as a group of automorphisms. It now follows from a theorem of Hecke [He1] that C_0 is isomorphic to the modular curve $X(11)$. Hecke's theorem asserts that the modular curve of prime level is determined up to isomorphism by its genus and its automorphism group. The genus of the modular curve of prime level p is well known to be

$$1 + \frac{(p-6)(p^2-1)}{24},$$

which is 26 for $p = 11$.

Definition (38.10): *We will refer to the singular locus C of the Hessian of f as the z-curve of the cubic threefold* Λ_f.

Remark (38.11): Since the rank of $M_x(f)$ is never less than 3 for a sufficiently generic cubic f, the null spaces of $M_x(f)$ as x varies over the z-curve C of Λ_f are the fibres of a rank 2 vector bundle on C. This bundle deserves further study, both for general f and in the case of Klein's cubic f_0, where it joins a number of other rank 2 vector bundles on the modular curve.

§39   Desingularization of the hessian

In the preceding section, we showed that condition P is an open condition and that Klein's cubic Λ_0 satisfies P. For the rest of this article, we will refer to cubics satisfying P as being **sufficiently generic**. In this section, we will prove the following result.

Theorem (39.1): *Let Λ be a sufficiently generic cubic threefold. Then the locus $\widehat{\mathcal{H}}(\Lambda)$ is nonsingular. In particular, the projection of $\widehat{\mathcal{H}}(\Lambda)$ onto either factor of $\mathbf{P}^4 \times \mathbf{P}^4$ is a desingularization of $\mathcal{H}(\Lambda)$.*

Proof: We first observe that our hypotheses imply that the rank ≤ 2 locus of $\mathcal{H}(\Lambda)$ is empty. For suppose x is a point of $\mathcal{H}(\Lambda)$, let k be the rank of $M_x(f)$ and suppose that $k \leq 2$. Then $\iota(x)$ has dimension $4 - k$, which is ≥ 2. First suppose that $k = 2$. Then $\iota(x)$ is a plane in $\mathbf{P}^4$. Every line λ in $\iota(x)$ must either meet the z-curve C in a divisor of degree 4 on λ or else it must lie entirely[21] in C. The latter can't occur since we have assumed that C is irreducible and has genus 26. It follows that the intersection of $\iota(x)$ with C is a plane quartic curve, which again contradicts the assumption that C is irreducible and has genus 26. In case $k = 1$, the argument is similar except that in this case we conclude that $\iota(x)$ meets C in a surface, contradicting the assumption that C is a curve.

---

[21] The lines through x in $\mathbf{P}^4$ form a $\mathbf{P}^3$ which we denote $\mathbf{P}^4/x$. The matrices $M_x(f)$ with $y \in \lambda$ have x in their kernels. Hence λ determines a pencil λ/x of quadrics in $\mathbf{P}^4/x$. Singular quadrics in $\mathbf{P}^4/x$ form a hypersurface of degree 4 in the projective space of all quadrics. Therefore, the pencil λ/x either has 4 singular quadrics, with multiplicity, or else consists entirely of singular quadrics. That immediately implies what we claimed for λ.

Since the rank ≤ 2 locus is empty and since Λ is nonsingular, we can apply the last lemma of §37. It therefore remains to prove that there do not exist 3 points x, y, z of C such that the lines $\iota(x)$, $\iota(y)$ and $\iota(z)$ form a triangle with vertices x, y, z. As noted above, the intersection of any $\iota(x)$ with x in C is a divisor of degree 4 on $\iota(x)$. This divisor can just as well be regarded as a divisor $\Xi(x)$ of degree 4 on C. If we associate to a point x of C the divisor $\Xi(x)$ on C, we define a correpondence which we denote Ξ. Because of the symmetry of the relation $y \in \iota(x)$, it follows that Ξ is a symmetric correspondence of bidegree $(4, 4)$.

We note that the existence of a triangle with vertices x, y, z and sides $\iota(x)$, $\iota(y)$ and $\iota(z)$ is equivalent to the existence of a fixed point for the threefold composition $\Xi \circ \Xi \circ \Xi$ of Ξ with itself. One can, in principle, count the number of fixed points using the Lefschetz fixed point formula for correspondences. For this, one would need to know the eigenvalues of Ξ acting on the cohomology of C. Now, as the cubic Λ varies among the open set of sufficiently generic cubics, the curve C and the correspondence Ξ vary with it. As they vary, the eigenvalues of Ξ vary continuously with Λ. On the other hand, since the correspondence Ξ must leave the rational cohomology of C invariant, the eigenvalues are algebraic numbers. The only way one can vary algebraic numbers continuously is for the numbers to remain constant. Therefore, we can compute the eigenvalues of Ξ by computing them in a special case. The special case that we will work with is the case Λ_0. In this case, the correspondence will be denoted Ξ_0.

I have shown in my articles [A2],[A4] that in this case Ξ_0 is the Hecke correspondence T_3. One can therefore find its eigenvalues either by looking them up in tables [Vé3] or be the geometric arguments given in [A2],[A4]. One finds that the eigenvalues of Ξ_0 on $H^1(C_0, \mathbf{R})$ are 2, with multiplicity 20, and -1 with multiplicity 32. By the Lefschetz fixed point formula, the number of fixed points of $\Xi \circ \Xi \circ \Xi$ in general is therefore

$$4^3 + 4^3 - 20 \cdot 2^3 - 32 \cdot (-1)^3 = 64 + 64 - 160 + 32 = 0.$$

This completes the proof of the theorem.

Theorem (39.2): *Let Λ be a sufficiently generic cubic threefold. Then the Jacobian variety of the z-curve of Λ decomposes into the sum of an abelian variety of dimension 10 and one of dimension 16.*

Proof: This follows at once from the determination of the eigenvalues, which are integers, of the correspondence Ξ described in the proof of the preceding theorem.

These abelian varieties will be studied further in [A15].

§40   Applications to mirror symmetry

In this section, we discuss some possible applications of this work to the phenomenon of mirror symmetry. While I don't propose to give a general definition of mirror symmetry, I will mention that a lot of recent work in this area is concerned with Calabi-Yau threefolds and the periods of integrals on these complex manifolds.

One example that has been quite closely studied is that of smooth quintic hypersurfaces in $\mathbf{P}^4$. Now, the hessian of a cubic threefold is a quintic hypersurface in $\mathbf{P}^4$, but it is certainly not smooth. On the other hand, there has been some recent interest in the boundary of the moduli space and it seems quite plausible that these hessians could arise in this connection. However, we will not explore that possibility directly. Instead, in this section we will show how the hessians of cubic threefolds can be more directly related to mirror symmetry.

As we have shown, the locus $\widehat{\mathcal{H}(\Lambda)}$ is a desingularization of the hessian $\mathcal{H}(\Lambda)$ of a sufficiently generic cubic threefold. Denote by h_1, h_2 the canonical generators of $H^2(\mathbf{P}^4 \times \mathbf{P}^4, \mathbf{Z})$. Since $\widehat{\mathcal{H}(\Lambda)}$ is defined by 5 bilinear equations in $\mathbf{P}^4 \times \mathbf{P}^4$, the first chern class of the normal bundle of $\widehat{\mathcal{H}(\Lambda)}$ is $5h_1 + 5h_2$. On the other hand, the first chern class of the tangent bundle of $\mathbf{P}^4 \times \mathbf{P}^4$ is also $5h_1 + 5h_2$. Therefore, the canonical bundle of $\widehat{\mathcal{H}(\Lambda)}$ has vanishing chern class. So $\widehat{\mathcal{H}(\Lambda)}$ is a Calabi-Yau manifold.

One can also use the results of this article to study the periods of $\widehat{\mathcal{H}(\Lambda)}$. Indeed, for every point x of $\mathcal{C}$, the fibre of $\widehat{\mathcal{H}(\Lambda)}$ over x is the projective line $\iota(x)$. We therefore have a natural mapping from the first homology group $H_1(\mathcal{C}, \mathbf{R})$ into the third homology group $H_3(\widehat{\mathcal{H}(\Lambda)}, \mathbf{R})$ which associates to any 1-cycle in $\mathcal{C}$ its preimage in $\widehat{\mathcal{H}(\Lambda)}$. This in turn induces a mapping of abelian varieties from the Jacobian variety $Jac(\mathcal{C})$ to the intermediate Jacobian $J_3(\widehat{\mathcal{H}(\Lambda)})$ of $\widehat{\mathcal{H}(\Lambda)}$ in the middle dimension.[22] Thus, we can learn about the periods of $\widehat{\mathcal{H}(\Lambda)}$ from its intermediate Jacobian and in turn from the Jacobian variety of $\mathcal{C}$. In the articles [A2], [A15], [A18], I discuss the properties of $Jac(\mathcal{C})$ in more detail and prove among other things that it is always the sum of two abelian varieties of dimensions 10 and 16. The question of the simplicity of these abelian varieties and of their possible relation to the intermediate jacobian variety of the cubic threefold was raised in [A2] and studied in more detail in [A15], [A18].

§41 The search for a quaternionic Hilbert modular threefold

Let $k = \mathbf{Q}(cos\ \frac{2\pi}{11})$ be the real subfield of the 11-th cyclotomic field. Then k is a totally real abelian quintic extension of $\mathbf{Q}$. Accordingly, it has 5 infinite places, denoted $\infty_0, \ldots, \infty_4$, all of them real. If S is a nonempty subset of the set of infinite places of k such that S has an even number of elements, denote by D_S the unique quaternion division algebra over k ramified at the places of S and at no other places of k. Denote by $\mathcal{O}_S$ a maximal order of D_S. Denote by $\mathcal{P}_S$ the unique prime of $\mathcal{O}_S$ lying over the prime 11 of $\mathbf{Q}$. Denote by Γ_S the group of units of norm 1 in $\mathcal{O}_S$ and by $\Gamma_S(\mathcal{P}_S)$ the subgroup of Γ_S consisting of elements congruent to 1 modulo $\mathcal{P}_S$. Denote by H_S the product of n copies of the upper half plane H, where $5 - n$ is the cardinality of S. Then $\Gamma_S(\mathcal{P}_S)$ acts on H_S with compact quotient and without fixed points. Denote the quotient X_S. Then X_S is a compact complex

---

[22] See Weil's article [W5] for the definition of the intermediate jacobian.

manifold of dimension n on which the group $G_S = \Gamma_s/\{\pm 1\}\Gamma(\mathcal{P}_S)$ acts. All these groups G_S may be identified with the group $G = PSL_2(\mathbf{F}_{11})$. If $n = 1$, one can show that the compact Riemann surface X_S has genus 26 and therefore, since G acts on it, is isomorphic to the modular curve $X(11)$. It follows, however, from the Shimizu formula [Shi1] that the representation of G on the space of holomorphic 3-forms on X_S when $n = 3$ is isomorphic to the representation of G on the space of holomorphic 1-forms on $X(11)$. Hence, as a G-module, the space of holomorphic 3-forms on X_S is the sum of irreducible representations of dimensions 5,10,11. Up to a scalar multiple, there is a unique basis $\phi_S^{(0)},\ldots,\phi_S^{(4)}$ for the 5-dimensional part such that the element $\begin{pmatrix} 1 & 1 \\ 0 & 1 \end{pmatrix}$ of G acts by the matrix

$$\begin{pmatrix} \zeta & 0 & 0 & 0 & 0 \\ 0 & \zeta^9 & 0 & 0 & 0 \\ 0 & 0 & \zeta^4 & 0 & 0 \\ 0 & 0 & 0 & \zeta^3 & 0 \\ 0 & 0 & 0 & 0 & \zeta^5 \end{pmatrix}$$

and the element $\begin{pmatrix} 2 & 0 \\ 0 & 6 \end{pmatrix}$ acts by the matrix

$$\begin{pmatrix} 0 & 0 & 0 & 0 & 1 \\ 1 & 0 & 0 & 0 & 0 \\ 0 & 1 & 0 & 0 & 0 \\ 0 & 0 & 1 & 0 & 0 \\ 0 & 0 & 0 & 1 & 0 \end{pmatrix}$$

with respect to this basis. Using that basis, we map X_S to $\mathbf{P}^4(\mathbf{C})$. Denote the image of X_S by Y_S. When $n = 1$, the image Y_S is Klein's degree 50 embedding of $X(11)$, the equations of which are given in [A11] The image Y_S when $n = 3$ is a locus which stands in a certain canonical G-invariant geometrical relationship to Klein's curve. That geometrical relationship between the Shimura varieties associated to different forms of the same arithmetic group remains to be articulated.

We don't even know how to prove that Y_S has dimension 3 when $n = 3$. Assuming it does have dimension 3, Y_S will be an irreducible hypersurface in $\mathbf{P}^4(\mathbf{C})$ invariant under G and will therefore be defined by an irreducible invariant of G. The full ring of invariants of G was determined in [A8],[A9]. A systematic study of the geometry of the invariant hypersurfaces and other invariant loci in $\mathbf{P}^4(\mathbf{C})$ was begun by Klein, pursued masterfully by Edge and explored further by the author. But we have only scratched the surface.

We do not know, assuming Y_S is a hypersurface when $n = 3$, whether the mapping of X_S to Y_S is birational. One way to test this would be to compare their birational invariants, such as the Todd genus of a desingularization. We know that the Todd genus of the smooth variety X_S is $1 - 26 = -25$, as it is for $X(11)$. Therefore we need to know the answer to the following question:

Question (41.1):   *Does there exist an irreducible G-invariant hypersurface Y in* $\mathbf{P}^4(\mathbf{C})$ *such that a desingularization $\tilde{Y}$ of Y has Todd genus -25?*

The simplest invariant is the cubic Λ_0, but its Todd genus is not -25. More generally, we have the following result.

Lemma (41.2):   *There is no nonsingular hypersurface in $\mathbf{P}^4(\mathbf{C})$ with Todd genus -25.*

Proof: We will use Hirzebruch's formula ([Hi2], §2.1) for the Todd genera on a complete intersection in a projective space. We will follow the presentation of [Sch1], Theorem 22.1.1, pp.159-160. Taking $n = 4$, $r = 1$, $a_1 = d$ and $y = 0$ in formula (2) of [Sch1],p.160, we find that the Euler characteristics of the invertible sheaf $\mathcal{O}_V(k)$ on a hypersurface of degree d in $\mathbf{P}^4(\mathbf{C})$ is given by

$$(41.3) \qquad \chi(V, \mathcal{O}_V(k)) = \text{coeff of } z^4 \text{ in } \frac{1}{(1-z)^{k+1}}(1 - (1-z)^d).$$

From the exact sequence

$$(41.4) \qquad 0 \to \mathcal{T}(V) \to j^*(\mathcal{T}(\mathcal{P}^4)) \to \mathcal{O}_V(d) \to 0,$$

where $\mathcal{T}$ denote the tangent sheaf, we have by the Riemann-Roch theorem,

$$(41.5) \qquad Td(V) = \chi(V, \mathcal{T}(V)) = \chi(V, j^*(\mathcal{T}(\mathbf{P}^4(\mathbf{C})))) - \chi(V, \mathcal{O}_V(d)),$$

where $Td(V)$ denotes the Todd genus of V. On the other hand, on $\mathbf{P}^4(\mathbf{C})$ we have

$$(41.6) \qquad 0 \to \mathcal{O} \to \mathcal{O}(1)^5 \to \mathcal{T}(\mathbf{P}^4(\mathbf{C})) \to 0,$$

where the middle term denotes the direct sum of 5 copies of $\mathcal{O}(1)$. Restricting (5.6) to V via j, equation (5.5) becomes

$$(41.7) \qquad Td(V) = 5\chi(V, \mathcal{O}_V(1)) - \chi(V, \mathcal{O}_V(0)) - \chi(V, \mathcal{O}_V(d)).$$

Therefore, by (5.3) we have
(41.8)

$$Td(V) = \text{coeff of } z^4 \text{ in } \left(5\frac{1}{(1-z)^2} - \frac{1}{(1-z)} - \frac{1}{(1-z)^{d+1}}\right)(1 - (1-z)^d)$$
$$= \text{coeff of } z^4 \text{ in } 5(1-z)^{-2} - 5(1-z)^{d-2} + (1-z)^{d-1} - (1-z)^{-d-1}.$$

Since the coefficient of z^4 in $(1-z)^k$, for any integer k, is the binomial coefficient $\binom{k}{4}$, the Todd genus $Td(V)$ of V is given by

$$Td(V) = 5\binom{-2}{4} - 5\binom{d-2}{4} + \binom{d-1}{4} - \binom{-d-1}{4}$$

$$(41.9) \qquad = 25 - 5\binom{d-2}{4} + \binom{d-1}{4} - \binom{d+4}{4}$$

$$= -\frac{5}{24}(d^4 - 10d^3 + 71d^2 - 134d).$$

If we now require $Td(V)$ to be -25, then d must satisfy the equation

$$(41.10) \qquad d^4 - 10d^3 + 71d^2 - 134d - 120 = 0.$$

Reducing the equation modulo 5, we see that d is divisible by 5 and we can write $d = 5e$, hence we have modulo 125 that

$$(41.11) \qquad\qquad -20e^2 - 9e + 1 \equiv 0.$$

Reducing this congruence modulo 5, we see that e is congruent to -1 modulo 5 and we can write $e = 5f - 1$. Therefore modulo 125 we have

$$(41.12) \qquad\qquad 30f - 10 \equiv 0,$$

which implies that f is congruent to 17 modulo 25 and we can write $f = 25g + 17$, hence

$$(41.13) \qquad d = 5e = 5(5f - 1) = 25f - 5 = 25(25g + 17) - 5 = 625g + 420.$$

In particular, d must have absolute value > 120. On the other hand, from Equation (41.10), the integer d must be a factor of 120, hence $|d| \leq 120$, which is a contradiction. This proves that there is no nonsingular hypersurface in $\mathbf{P}^4(\mathbf{C})$ with Todd genus -25.

The next invariant after Klein's cubic Λ_0 is the Hessian $\mathcal{H}_0$ of Λ_0. From the explicit desingularization $\widehat{\mathcal{H}}_0$ of $\mathcal{H}_0$ considered here, we know that the Todd genus of $\mathcal{H}$ is 0, not -25. In view of the delicacy of the explicit desingularization, it is greatly to be hoped that other methods of computing the Todd genus of a desingularization, such as intersection homology and L^2-cohomology, will prove effective. In this connection, a close examination of the case of the Hessian, considered in this paper, from these other points of view would be very instructive.

§42   Techniques of Gordan and Noether

We now prove the following lemma using techniques from the classical article [G-N] of Paul Gordan and Max Noether. We really do not even scratch the surface of their beautiful and ingenious paper here since it is really not pertinent to the goals of the present article. However, we plan to give a full exposition of their results in another article.

Theorem (42.1):   *The hessian of a nonsingular form of degree $d \geq 2$ in $n + 1$ variables does not vanish identically.*

Proof: Let F be a nonsingular form of degree $d \geq 2$ in $x_0, \ldots, x_n$. For $0 \leq i \leq n$, denote by F_i the partial derivative of F with respect to x_i. Since the F_i are homogeneous of degree $d - 1$, they are the coordinates of a homogeneous mapping from $\mathbf{C}^{n+1}$ to itself. That homogeneous mapping induces a rational mapping, which we denote $[dF]$ from $\mathbf{P}^n(\mathbf{C})$ to itself. If the Hessian of F is identically 0 then there is a polynomial relation among the partials of F_i. Let $\Pi = \Pi(y_0, \ldots, y_n)$ be a nonzero homogeneous polynomial of minimal degree in $n + 1$ variables $y_0, \ldots, y_n$ such that

$$(42.2) \qquad\qquad \Pi(F_0, \ldots, F_n) = 0.$$

For $0 \leq i \leq n$, denote by Π_i the derivative of Π with respect to y_i and define π_i by

$$\pi_i = \Pi_i(F_0, \ldots, F_n).$$

If m is the degree of Π, we have

(42.3)
$$\sum_{i=0}^{n} y_i \Pi_i = m\Pi,$$

by a theorem of Euler. Because Π was chosen to have minimal degree, for some i we have

$$\pi_i \neq 0.$$

Replacing y_i by F_i in (42.3) for $0 \leq i \leq n$, we have

$$0 = m\Pi(F_0, \ldots, F_m)$$
$$= \sum_{i=0}^{n} F_i \Pi(F_0, \ldots, F_n)$$
$$= \sum_{i=0}^{n} F_i \pi_i.$$

Denoting by D the derivation

$$D = \sum_{i=0}^{n} \pi_i \frac{\partial}{\partial x_i},$$

we can write this last identity as

$$DF = 0.$$

We will say that a polynomial or rational function annihilated by D is D-**constant**. Thus F is a D-constant. If we take the derivative of the relation

$$\Pi(F_0, \ldots, F_n) = 0$$

with respect to x_j, it follows that F_j is also D-constant. Indeed, we have

$$0 = \frac{\partial}{\partial x_j} \Pi(F_0, \ldots, F_n)$$
$$= \sum_{i=0}^{n} \pi_i \frac{\partial F_i}{\partial x_j}$$
$$= \sum_{i=0}^{n} \pi_i \frac{\partial F_j}{\partial x_i},$$

so $DF_j = 0$. Since F has degree $d \geq 2$ and since we have

$$(d-1)F_j(\pi_0, \ldots, \pi_n) = \sum_{i=0}^{n} \pi_i \frac{\partial F_j}{\partial x_i} = DF_j = 0$$

where the π_i are not identically 0, it follows that the polynomials F_j have a common zero, so F is singular.

Appendix V

New Abelian Varieties Associated to Cubic Threefolds
by Allan Adler

§43 Introduction

It is well known how to associate an abelian variety to a nonsingular cubic three-fold Λ. One simply considers the intermediate Jacobian variety $J_3(\Lambda)$ of the cubic threefold in the middle dimension. It is known that $J_3(\Lambda)$ is a principally polarized abelian variety of dimension 5 which contains a lot of information about Λ. Indeed, according to the Torelli theorem of Clemens and Griffiths [C-G], one can recover Λ from $J_3(\Lambda)$.

In this article, we show how to associate some new abelian varieties to a sufficiently generic cubic threefold. Our constructions are motivated by a close examination of the special case of Klein's cubic threefold

$$v^2w + w^2x + x^2y + y^2z + z^2v = 0,$$

which has a number of remarkable features. Two of the most useful for our purposes are:
(1) its invariance under a group of collineations isomorphic to $PSL_2(\mathbf{F}_{11})$ and
(2) the fact, discovered by Felix Klein, that the singular locus of the Hessian of Klein's cubic is a smooth curve isomorphic to the modular curve $X(11)$ of order 11.

We give two essentially different constructions which associate new abelian varieties to a cubic threefold. The first construction associates to a generic cubic threefold Λ an abelian variety $A_{10}(\Lambda)$ of dimension 10 and an abelian variety A_{16} of dimension 16. The first construction was first discovered in the author's unpublished paper [A2]. The argument is reproduced, with applications to mirror symmetry, in the author's paper [A16]. So the first construction will only be summarized here and the reader is referred to [A16] for details of the proof. One point left open in [A2] was the question of the simplicity of these two abelian varieties. It was hoped that the 16 dimensional abelian variety would decompose into the sum of abelian subvarieties of dimensions 5 and 11 and that the 5 dimensional component would be isogeneous to the intermediate Jacobian of Λ. These abelian varieties are simple in general. This is proved in §46. Originally all the author knew how to do was to prove that the resulting abelian schemes of dimensions 10 and 16 were simple over the space of sufficiently generic cubic threefolds. However, Torsten Ekedahl showed how to modify the author's original argument to show that these abelian schemes are in fact absolutely simple. Both arguments are presented in §46.

The second construction associates to each generic cubic threefold Λ a five-dimensional family of abelian varieties of dimension 21. It turns out that, unlike the first, this second construction is unexpectedly related to the intermediate Jacobian of Λ.

These two constructions are complementary to each other in a sense which sheds some light on the well-known decomposition [Hu1],[He2] of the Jacobian variety $J(X(11))$ into $PSL(2,11)$ invariant abelian subvarieties of dimensions 5, 10 and 11. It follows from our investigations that in the moduli space of curves of genus 26, there are two subvarieties $\mathcal{H}esse$ and $\mathcal{P}faff$ whose intersection contains the point of the moduli space corresponding to $X(11)$. The modular curve $X(11)$, as it occurs in the family $\mathcal{H}ess$, is Klein's z-curve of level 11, while as it occurs in the family $\mathbf{P}faff$, it is Klein's A-curve of level 11. Therefore our constructions may be regarded as generalizing the notion of z-curve and A-curve to the general context of cubic threefolds.

The curves of the family $\mathcal{H}esse$ have the property that their Jacobian varieties are isogeneous to the sum of an abelian variety of dimension 10 and one of dimension 16. The curves of the family $\mathcal{P}faff$ have the property that their Jacobian varieties are isogeneous to the sum of an abelian variety of dimension 5 and one of dimension 21. Thus, the decomposition

$$26 = 5 + 10 + 11$$

for the modular curve $X(11)$ is a specialization of the behavior of both families, since

$$5 + 21 = 5 + (10 + 11) = 10 + (5 + 11) = 10 + 16.$$

We don't know whether the 21 dimensional abelian varieties one obtains are simple in general.

Our investigations also shed light on the intermediate Jacobian $J_3(\Lambda)$ of a cubic threefold. Indeed, we show that the curves of the second family which arise from a given cubic threefold are such that the 5 dimensional component of their Jacobian varieties are all isogeneous to $J_3(\Lambda)$.

The reason for the names $\mathcal{H}esse$ and $\mathcal{P}faff$ is as follows: the family $\mathcal{H}esse$ is the family of singular loci of the Hessians of cubic threefolds. For a generic cubic, these are smooth curves of genus 26 whose Jacobian varieties are isogenous to the sum of an abelian variety of dimension 10 and one of dimension 16. The family $\mathcal{P}faff$ is constructed using the fact, presented here apparently for the first time, that a generic cubic form in 5 variables is expressible in ∞^5 essentially different ways as the Pfaffian of a skew-symmetric 6×6 matrix whose entries are linear forms in 5 variables. Given such a representation of the cubic, each point of the cubic determines a projective line in $\mathbf{P}^5$ and the family of ∞^3 lines so obtained sweeps out a quartic fourfold in $\mathbf{P}^5$ whose singular locus consists of a smooth curve of degree 25 and genus 26. The Jacobian variety of this curve is the sum of an abelian variety of dimension 21 and the an abelian variety isogenous to the intermediate Jacobian of the cubic threefold.

The discovery of this new canonical form for cubic threefolds has important applications to the study of rank 2 vector bundles on cubic threefolds. In effect, we have constructed a natural 5 dimensional family of rank 2 vector bundles on a generic cubic threefold. Since the cubic threefold is the union of lines, there being 6 lines through a generic point of the threefold and lying in it, one can also inquire into the jumping lines for each of these rank 2 vector bundles. We are able to describe these jumping lines in terms of the relevant geometry and this, in turn, is what allows us to relate this second construction to the intermediate Jacobian of the cubic threefold.

The proof that the abelian varieties $A_{10}(\Lambda)$ and $A_{16}(\Lambda)$ are simple, in the technical sense mentioned above, relies on the unexpected fact, discovered in [A13], that in characteristic 3 the modular curve $X(11)$ admits the Mathieu group M_{11} as an automorphism group.

In connection with these investigations, we would also like to draw attention to the desirability of computing the ring of invariants of quinary cubics. Indeed, two cubic threefolds in $\mathbf{P}^4(\mathbf{C})$ have isomorphic intermediate Jacobians if and only if one can be carried to the other by a collineation of $\mathbf{P}^4(\mathbf{C})$. Hence, the intermediate Jacobians of cubic threefolds are a complete set of invariants of quinary cubics. We therefore have every right to ask for a theory of cubic threefolds which unifies both the point of view of invariant theory and the point of view of the intermediate Jacobian variety, just as we possess such a theory for ternary cubics.

The structure of this article is as follows. In §44, we summarize the first construction, explain the technical notion of simplicity that we are using and give an informal sketch of the argument which we use to prove that the abelian varieties A_{10} and A_{16} are simple, in our technical sense, for a generic cubic threefold. We feel that this informal sketch will be helpful to the reader in advance of plunging into the technical details of the scheme theoretic proof which is given in §46. Indeed, until the author mastered enough technical details to construct the arguments of §46, all he possessed was the sketch in §44, without which the argument of §46 might never have appeared. In §45, we recall technical results from [EGA], [FGA] and [BLR] which we use in the arguments of §46. In §47, present the canonical form for the quinary cubic as a Pfaffian. In §48 we use this canonical form to construct rank 2 vector bundles on the cubic threefold. Since Klein's embedding of the modular curve $X(11)$ as a curve of degree 50 in $\mathbf{P}^4$ maps the curve into Klein's cubic threefold, this also gives us a family of rank 2 vector bundles on the modular curve. In §48 we study the locus of jumping lines of the vector bundles we have constructed on cubic threefolds. In §50, we show how to map the associated $\mathbf{P}^1$ bundles of these vector bundles birationally onto a quartic hypersurface in $\mathbf{P}^5$. By examining the singular locus of this quartic hypersurface, which turns out to be the unique quartic invariant for $SL_2(\mathbf{F}_{11})$ in the case of (47.1), we give the equations of Klein's A-curve of level 11 for the first time and thereby generalize the notion of A-curve to the context of cubic threefolds. For it turns out that the singular locus of the quartic consists of an smooth irreducible curve of degree 25 and genus 26 for a sufficiently generic Pfaffian. In §52, we show how to decompose the Jacobian variety of one of these generalized A-curves as the sum of an abelian variety of dimension 21 and one of dimension 5, the latter being isogenous to the intermediate Jacobian of the cubic threefold (isomorphic in the case of 47.1). In §53, we consider the possibility of generalizing the fundamental intertwining operator of [A-R], §20, to the context in which we find our generalized A-curves and z-curves.

In many places in this article, we use the terminology **sufficiently generic** to describe an object, usually an element of $\mathcal{P}$, lying in a nonempty Zariski open set which depends on the particular situation. We are constantly concerned with proving that the predicate "sufficiently generic" applies, in these situations, to the example (47.1), i.e. that the example (47.1) lies in the nonempty Zariski open set. Often that is the only way we know that the set is in fact nonempty. The terminology "sufficiently generic" was used in [A16] with a more precise meaning appropriate to the study of generalized z-curves (but not generalized A-curves). There we kept careful track of the nonempty Zariski open sets that arose and their nesting. We could have followed the same approach here, and there would have been some advantages in doing so, but not enough really to justify modifying the article to facilitate such an approach.

I am grateful to Yum-Tong Siu, Torsten Ekedahl and Gabor Megyesi for their help in proving that the singular locus of the quartic $\Sigma(M)$ is a curve. Torsten Ekedahl also kindly read portions of the manuscript and suggested a number of valuable improvements. I am also indebted to Ofer Gabber for patiently listening to my idea for the main argument of §46 and for reassuring me that it was not complete nonsense.

§44   The idea behind the proof of simplicity

Before plunging into the technical details of the proof of Theorem 46.12, I would like to present a sketch of the basic idea in the intuitive form in which it originally occurred to me and without reference to any notions of schemes.

Over the complex numbers, denote by U the set of nonsingular cubic hypersurfaces in $\mathbf{P}^4$ such that the singular locus of the Hessian of the cubic is a smooth curve of degree 20 and genus 26. Then U is Zariski open in the projective space of all cubic hypersurfaces. The curves of genus 26 one obtains in this way form a family of curves indexed by U. The set U is nonempty because one shows that it contains Klein's cubic threefold

$$v^2w + w^2x + x^2y + y^2z + z^2v = 0$$

for which the corresponding curve is known to be the modular curve $X(11)$. The singular locus of the Hessian is also the rank 3 locus of the matrix of second partials [A13], [A16]. If x is a point on the curve $\mathcal{C}$ associated to a cubic threefold Λ in U then the null space of the matrix of second partials is 2 dimensional and therefore determines a projective line $\iota(x)$ in $\mathbf{P}^4$. That line meets the curve in 4 points, with suitably defined multiplicity. The mapping Ξ that associates to x those 4 points where $\iota(x)$ meets $\mathcal{C}$ is a symmetric $(4,4)$ correspondence on the curve $\mathcal{C}$. That correspondence then acts on the Jacobian variety $Jac(\mathcal{C})$ of $\mathcal{C}$.

One can determine the eigenvalues of Ξ on the tangent space to the identity of $Jac(\mathcal{C})$ in the following way. We know that the eigenvalues have to vary continuously as one varies the cubic Λ in U. On the other hand, Ξ leaves invariant the rational cohomology of $\mathcal{C}$, so the eigenvalues are algebraic integers. Therefore they remain constant under continuous variation. So it is enough to determine them in the special case of Klein's cubic. But in that case, $\mathcal{C}$ is the modular curve $X(11)$ and I have proved that the correspondence Ξ is the Hecke correspondence T_3, whose eigenvalues on the tangent space to the identity of $Jac(X(11))$ are well known to be -1, with multiplicity 16, and 2, with multiplicity 10. The same is therefore true for any cubic Λ in U. If we denote by A_{10} the connected component of the identity in the abelian subvariety of $Jac(\mathcal{C})$ consisting of points sent to their inverses by Ξ, we therefore get an abelian variety of dimension 10. Similarly, if we consider points x sent to their doubles $2x$ by Ξ, we get an abelian variety A_{16} of dimension 16. Furthermore, we have a decomposition of $Jac(\mathcal{C})$ as the sum of A_{10} and A_{16}.

This is the situation for every Λ in U. It is natural to ask whether these abelian varieties can be further decomposed. As a first step in answering this, we first note that U, the family of curves $\mathcal{C}$, the family of their Jacobian varieties $Jac(\mathcal{C})$ and the family of endomorphisms Ξ admit an action by $PGL_5(\mathbf{C})$. In particular, any generic behavior of these abelian varieties must specialize, in the case of Klein's cubic, to behavior that is invariant under the isotropy group of Klein's cubic, i.e. $PSL_2(\mathbf{F}_{11})$. Now, the action of $PSL_2(\mathbf{F}_{11})$ on holomorphic 1-forms on $X(11)$ decomposes into irreducible representations of degrees 5, 10 and 11. This reflects the fact that the Jacobian variety $Jac(X(11))$ accordingly decomposes into abelian varieties of these dimensions which are invariant under the group. Furthermore, the Jacobian has no further equivariant decomposition (although $Jac(X(11))$ is known to be isogenous to the product of 26 elliptic curves). It follows that for generic Λ, the abelian variety A_{10} is simple over the function field of U and that the abelian variety A_{16} is either simple over the function field of U or else is the sum of an abelian variety A_5 of dimension 5 and an abelian variety A_{11} of dimension 11. We will show that in fact, A_{16} is simple over the function field of U for generic Λ.

Suppose on the contrary that A_{16} decomposes as the sum of A_5 and A_{11}. Let T be an endomorphism of $Jac(\mathcal{C})$ whose image is A_5. By varying Λ, we obtain a family of such endomorphisms. Since the endomorphism ring of the family $Jac(\mathcal{C})$ is discrete and since it is normalized by the action of $PGL_5(\mathbf{C})$, it follows that $PGL_5(\mathbf{C})$ commutes with T. In particular, in the case of Klein's cubic, T commutes with $PSL_2(\mathbf{F}_{11})$ and its image is a 5 dimensional abelian variety invariant under the group. This we already knew. But one can show (cf. [A7]) that any 5 dimensional abelian variety on which $PSL_2(\mathbf{F}_{11})$ acts nontrivially must be the product of 5 copies of an elliptic curve E with complex multiplication by $\mathbf{Q}(\sqrt{-11})$.

The discreteness of the endomorphism ring of $Jac(\mathcal{C})$ implies that T is defined over a number field k. We now reduce the entire situation modulo a prime lying over 3 in k. In characteristic 3, the Hessian of a cubic form doesn't really depend on the cubic. Instead, it depends on the cubic modulo cubes of linear forms, since the derivative of a cube is 0 in characteristic 3. Instead of the space of all cubic forms, consider the space of all cubics modulo cubes and form the corresponding projective space. Inside it, there will be an open set, denoted U' consisting of those cubics modulo cubes for which the rank 3 locus of the Hessian is a smooth curve of genus 26. The set U' is nonempty since it contains the reduction of Klein's cubic modulo 3. We obtain the analogous family of curves and their Jacobians. We also obtain the reduction of the correspondence T. Now, although the Hessians, the curves and their Jacobians only depend on the cubic modulo cubes, one has to prove that T also only depends on the cubic modulo cubes. But that follows from the fact that any rational mapping from affine space into an abelian variety is constant. So we also have the endomorphism T in this setting. The entire construction is equivariant for the action of the algebraic group PGL_5 over $\mathbf{F}_3$. Therefore, the endomorphism T, in the case of Klein's cubic, commutes with the subgroup of PGL_5 preserving Klein's cubic modulo cubes.

I have shown in [A13] that although the subgroup preserving Klein's cubic still contains $PSL_2(\mathbf{F}_{11})$ in characteristic 3, a larger group preserves it modulo cubes of linear forms, namely the Mathieu group M_{11}. In particular, M_{11} must act on the image of T. But the image of T in characteristic 3 will be reduction modulo 3 of the image of T in characteristic 0, hence is the reduction of the product of 5 copies of E. We therefore get a nontrivial homomorphism of M_{11} into $GL_5(End(E))$. Since the prime 3 splits in the quadratic field $\mathbf{Q}(\sqrt{-11})$, the endomorphism ring of E in characteristic 3 is still the ring of integers of the quadratic field. In particular, $End(E)$ is a subring of $\mathbf{C}$ and we obtain a nontrivial complex representation of M_{11} of degree 5. But it is well known that M_{11} has no such complex representation and that is a contradiction. That proves that A_{16} is generically simple.

Although I found these arguments intuitively satisfying and convincing, I didn't feel that they constituted a completely rigorous proof. So I rewrote the argument more carefully using the language of schemes. We will present that scheme theoretic argument in §46.

§45   Standard Results About Schemes

This section contains a number of technical results from [EGA] and [BLR] which we will need in §46. We quote them here for the convenience of the reader. Their collection in this section offers little in the way of exposition. The reader is advised to skip or, at best, only to scan this section on first reading and refer back to it when going through the proofs in §46. For genuine exposition, it is best to read the ambient material in [EGA] and [BLR].

Lemma   (45.1):   (Chevalley) *Let X, Y be schemes and let $f : X \to Y$ be a morphism which is locally of finite type. For every integer n, the set $F_n(X)$ of all $x \in X$ such that*

$$dim_x(f^{-1}(f(x))) \geq n$$

is closed. In other words, the function

$$x \mapsto dim_x(f^{-1}(f(x)))$$

is upper semi-continuous on X.

Proof: This is just EGA IV$_3$.13.1.3.

Definition (45.2): (EGA IV$_3$.14.1.1) *A continuous mapping* $\psi : X \to Y$ *between topological spaces is said to be* **open** *at a point* x *if every neighborhood of* x *is mapped by* ψ *to a neighborhood of the point* $\psi(x)$ *in* Y.

Definition (45.3): (EGA IV$_3$.14.3.3) *Let* $f : X \to Y$ *be a morphism of schemes and let* x *be an element of* X. *We say that* f *is* **universally open** *at* x *if for every* Y-*scheme* $g : Y' \to Y$, *the morphism* $f' = f_{(Y')} : X' \to Y'$ *is open at every point of* x' *of* X' *whose image in* X *is* x. *Here* X' *is the fibre product of* X *and* Y' *over* Y.

Lemma (45.4): *Let* X *be a scheme, let* Y *be a locally noetherian scheme and let* $f : X \to Y$ *be a morphism which is locally of finite type. Let* $\mathcal{F}$ *be a coherent* $\mathcal{O}_X$-*Module, let* x *be a point of* X *and let* $y = f(x)$. *If the following four conditions hold then* $\mathcal{F}$ *is* f-*flat at* x:
  (i) $Supp(\mathcal{F}) = X$;
  (ii) f *is universally open at the generic points of the irreducible components of* $f^{-1}(y)$ *containing* x;
  (iii) $\mathcal{F}_y = \mathcal{F} \otimes_{\mathcal{O}_Y} \kappa(y)$ *is geometrically reduced at the point* x;
  (iv) *The ring* $\mathcal{O}_y$ *is reduced.*

Proof: This is just EGA IV$_3$.15.2.2.

Definition (45.5): (EGA IV$_1$ 0.23.2.1)*If* A *is a ring, we denote by* A_{red} *the quotient of* A *by its nilradical. We will say that a local ring* A *is* **unibranch** *if the ring* A_{red} *is entire and if the integral closure* A'_{red} *of* A_{red} *is also a local ring. We will say that* A *is* **geometrically unibranch** *if it is unibranch and if the residue field of the local ring* A'_{red} *is a radical extension of the residue field of* A.

Since we will always be working with irreducible schemes, we use the following definition instead of the more general version given in EGA IV$_3$14.3.2.

Definition (45.6): (EGA IV$_3$14.2.2) *Let* X, Y *be irreducible schemes and let* $f : X \to Y$ *be a dominant morphism which is locally of finite type. Let* η *be the generic point of* Y. *We say that* f *is* **equidimensional** *at a point* x *if*

$$dim_x(f^{-1}(f(x))) = dim(f^{-1}(f(\eta))).$$

Lemma (45.7): (**Chevalley's criterion for universal openness**) *Let* X, Y *be schemes and let* $f : X \to Y$ *be a morphism locally of finite type.*
  (i) *If* f *is equidimensional at a point* $x \in X$ *and if* $y = f(x)$ *is a point where* Y *is geometrically unibranch then* f *is universally open at the point* x.
  (ii) *If* Y *is geometrically unibranch then* f *is universally open at all of the points where* f *is equidimensional and the set of such points is open in* X. *In particular, if* f *is equidimensional it is universally open.*

Lemma (45.8): *Let* $f : X \to Y$ *be a morphism which is proper, flat and of finite presentation. For* $y \in Y$, *let* X_y *denote the fibre of* f *over* y. *The following subsets of* Y *are then open:*
  (i) *The set of* $y \in Y$ *such that* X_y *is geometrically regular.*

(ii) The set of $y \in Y$ such that X_y is geometrically entire.

Proof: These are both stated in EGA IV$_3$.12.2.4. The assertion (i) is EGA IV$_3$ 12.2.4(iii) and the assertion (ii) is EGA IV$_3$.12.2.4(viii).

Lemma (45.9): *Let X, Y be schemes and suppose that Y is locally noetherian. Let $f : X \to Y$ be a flat proper morphism. For each nonnegative integer n, denote by $\mathbf{R}^n f_*(\mathcal{O}_X)$ the n-th direct image sheaf of $\mathcal{O}_X$ under f and let the rank of the stalk of $\mathbf{R}^n f_*(\mathcal{O}_X)$ over a point y of Y be denoted $h_n(y)$, the rank being taken as a module over the residue field $\kappa(y)$ of the local ring of y. Then the integer valued function χ defined on Y by*

$$\chi(y) = \sum_n (-1)^n h_n(y)$$

is locally constant on Y.

Proof: This is a special case of EGA III$_2$.7.9.4. There, one only requires the morphism f to be proper. The structure sheaf $\mathcal{O}_X$ is replaced by a complex $\mathcal{P}$. of coherent Y-flat modules. Since we can identify $\mathcal{O}_X$ with a complex whose only nonzero term is in degree 0, saying that the complex is Y-flat amounts to saying that f is flat, which is why we added that hypothesis here. The function χ is still the Euler characteristic but the sheaf $\mathbf{R}^n f_*(\mathcal{O}_X)$ is replaced by the hypercohomology of the complex $\mathcal{P}$. with respect to the morphism f. But for a complex concentrated in degree 0, the n-hypercohomology is easily seen from the definition (EGA III$_2$.6.2.1) to be the same as the n-direct image sheaf.

We also need some facts about the relative Picard functors $Pic_{X/S}$ and $Pic^0_{X/S}$. Since Grothendieck's [FGA] was not available to us when this article was written, we relied on our copy of the book [BLR]. Hence we follow their discussion.

Definition ([BLR], §8.1, p.199) (45.10): *Let S be a scheme and let (Sch/S) denote the category of S-schemes. Let $\mathcal{M}$ be a class of morphisms of (Sch/S) which contains all isomorphisms and is closed under fibre products and under composition of composable morphisms. A contravariant functor from (Sch/S) to the category $(Sets)$ of sets is also called a **presheaf**. A presheaf is said to be a **sheaf with respect to $\mathcal{M}$** or a **$\mathcal{M}$-sheaf** if the following two conditions are satisfied: (1) if $(T_i)_{i \in I}$ is any family of S-schemes then the canonical mapping*

$$F\left(\coprod_{i \in I} \right) \to \prod_{i \in I} F(T_i)$$

is an isomorphism. (2) if $T' \to T$ is a morphism in $\mathcal{M}$ then the sequence

$$F(T) \to F(T') \rightrightarrows F(T'')$$

is exact, where $T'' = T' \times_T T'$ and where the double arrows on the right are induced by the two projections of T'' onto T'.

The classes with which one is concerned in the definition of the relative Picard functor are the following:

(i) The class $\mathcal{M}_{Zar}$ of all morphisms in (Sch/S) of the form

$$\coprod_{i \in I} T_i \to T$$

where the morphisms $T_i \to T$ are open immersions whose images form an open covering of T. A sheaf with respect to $\mathcal{M}_{Zar}$ is called a **Zariski sheaf**.

(ii) The class $\mathcal{M}_{fppf}$ of faithfully flat finitely presented morphisms of S-schemes. A presheaf is said to be a **fppf-sheaf** if it is both a Zariski sheaf and a sheaf with respect to $\mathcal{M}_{fppf}$.

For every scheme Y, we define $Pic(Y)$ to be $H^1(Y, \mathcal{O}_Y^*)$. For every S-scheme X, we define the presheaf $P_{X/S}$ on (Sch/S) by the rule

$$P_{X/S}(T) = Pic(X \times_S T).$$

The functor $P_{X/S}$ is not a Zariski sheaf ([BLR],pp.200-201) because a locally trivial line bundle is not globally trivial in general. However, [BLR], p.201, gives a construction for the sheafification of a presheaf F with respect to a class $\mathcal{M}$ of morphisms. Such a sheafification is by definition a morphism of presheaves $\nu : F \to \tilde{F}$, where $\tilde{F}$ is a sheaf with respect to $\mathcal{M}$, such that any morphism from F to an $\mathcal{M}$-sheaf factors through ν. One constructs $\tilde{F}$ and ν by using descent data. The explicit construction is summarized in the following proposition, extracted from the discussion on p.201 of [BLR].

Proposition (45.11):   Let F be a presheaf on (Sch/S) and let $\mathcal{M}$ be a class of morhisms of (Sch/S) which contains all isomorphisms, is closed under fibre products and closed under composition of composable morphisms. For every morphism $T' \to T$ in $\mathcal{M}$, let $T'' = T' \times_T T'$ and denote by $H(T'/T, F)$ the subset of $F(T')$ consisting of all elements ξ of $F(T')$ such that for some T''-scheme $\tilde{T}$ the two images of ξ in $F(\tilde{F})$ in the diagram

$$F(T') \rightrightarrows F(T'') \to F(\tilde{T})$$

coincide. If the class $\mathcal{M}$ is not too big (e.g. if the skeleton of the class of T-schemes in $\mathcal{M}$ is a set for all S-schemes T) then the sets $H(T'/T, F)$ form an inductive system of sets whose direct limit we denote $\tilde{F}(T)$. Then the assignment $T \to \tilde{F}(T)$ defines a sheaf of sets on (Sch/S). Furthermore, the canonical morphism from F to $\tilde{F}$ is a sheafification of F.

The construction of the relative Picard functor is then contained in the following proposition, which we have extracted from the discussion on p.201 of [BLR].

Proposition (45.12):   Denote by $P'_{X/S}$ the sheafification of $P_{X/S}$ with respect to the class $\mathcal{M}_{fppf}$. Then the sheafification of $P'_{X/S}$ with respect to the class $\mathcal{M}_{Zar}$ is an fppf-sheaf and is denoted $Pic_{X/S}$.

Definition (45.13):   The fppf-sheaf $Pic_{X/S}$ is called the **relative Picard sheaf** of X/S.

The following lemma will be needed for the arguments of §46. It is not stated in [BLR] so I have provided my own proof. I'm informed that it may also be found in

[FGA] but I haven't seen it myself. In order to facilitate the discussion we introduce some notation. If $f : S' \to S$ is a morphism of schemes, we denote by f_* the functor $(Sch/S) \to (Sch/S')$ defined by[23] $f_*(g) = f \circ g$ for every S' scheme $g : U' \to S'$. We will also write $f_*(g)$ as $f_*(U)$. If X is an S-scheme, we denote by $f^*(X)$ the S'-scheme $X \times_S S'$. If $F : (Sch/S) \to (Sets)$ is a presheaf, we denote by $f^*(F)$ the presheaf $F \circ f_*$ on (Sch/S').

Lemma (45.14):   *The functors $P_{X/S}$, $P'_{X/S}$ and $Pic_{X/S}$ all commute with base change. In other words, if $f : S' \to S$ is a morphism of schemes and if X is an S-scheme then we have*

(1) $f^*(P_{X/S}) = P_{Y/S'}$
(2) $f^*(P'_{X/S}) = P'_{Y/S'}$
(3) $f^*(Pic_{X/S}) = Pic_{Y/S'}$,

where $Y = f^*(X)$.

Proof: Let $g : U' \to S'$ be an S'-scheme. Then we have

$$P_{Y/S}(U') = Pic(U' \times_{S'} Y) = Pic(U' \times_{S'} S' \times_S X)$$
$$= Pic(f_*(U') \times_S X) = (P_{X/S}(f_*(U'))) = f^* P_{X/S}(U'),$$

which proves (1). Next let V be an S'-scheme. We make the simple observation that the class of S'-morphisms with target V coincides with the class of S-morphisms with target f_*V, since both coincide with the class of fppf V-schemes. Therefore the index sets used in computing the direct limits defining $P'_{Y/S'}(V)$ and $P'_{X/S}(f_*V)$ coincide. Furthermore, under the natural identification (1), for any fppf morphism $U \to V$ of S'-schemes we have

$$H(U/V, P_{Y/S}) = H(f_*U/f_*V, P_{X/S}).$$

This follows from the fact that the computation of both of these groups is really done in the category of V-schemes, where the two computations coincide. This proves (2). The same argument shows that (2) implies (3).

In case S is the spectrum of a field, the relative Picard functor $Pic_{X/S}$ is representable by a group scheme ([FGA],#236, Thm.2.1, [BLR], Thm.8.4/3). In that case the connected component of the identity of $Pic_{X/S}$ is denoted $Pic^0_{X/S}$. For a general scheme S, $Pic^0_{X/S}$ is defined ([BLR],p.233) to be the subfunctor of $Pic_{X/S}$ consisting of all elements whose restrictions to the fibres X_s of X, for $s \in S$, lie in $Pic^0_{X_s/\kappa(s)}$.

Corollary (45.15):   *The functor $Pic^0_{X/S}$ commutes with base change. In other words, if $f : S' \to S$ is a morphism of schemes and $Y = f^*X$ then for every S'-scheme U we have*

$$Pic^0_{Y/S}(U) = Pic^0_{X/S}(f_*U).$$

Proof: Since $Pic^0_{X/S}$ is defined fibrewise and since $Pic_{X/S}$ commutes with base change, we may assume that S is the spectrum of a field K and that S' is the

---

[23]   This should not be confused with direct image sheaves, which uses a similar notation.

spectrum of an extension field K' of K. In this case, $Pic_{X/S}$ is a group scheme over K which, when viewed as a group scheme over K' is simply $Pic_{Y/S}$. According to [Ga1], Lemma 2.1.2 (cf.EGA IV.4.5.8, IV.4.5.14), the product of two connected schemes over a field K is connected provided that at least one of them has a rational point over that field. Therefore, taking one of the schemes to be $Pic^0_{X/S}$ and the other to be $Spec(K')$, we see that $Pic^0_{X/S}$ remains connected over K', hence we are done.

The following lemma occurs as Theorem 9.3/7 of [BLR], p.258.

Lemma (45.16): *Let S be a normal strictly local scheme and let $f : X \to S$ be a flat projective morphism whose geometric fibres are reduced and connected curves. Then the Jacobian of X is a smooth and separated S-scheme. It coincides with $Pic^0_{X/S}$ as defined in [BLR] 8.4.*

The following lemma occurs as Proposition 9.4/4 of [BLR], p.260.

Lemma (45.17): *Let $f : X \to S$ be a proper smooth morphism of schemes whose geometric fibres are connected curves. Then $Pic^0_{X/S}$ is an abelian S-scheme and there is a canonical S-ample rigidified line bundle $\mathcal{L}(X/S)$ on $Pic^0_{X/S}$.*

The following four lemmas are taken from [SGA1].

Lemma ([SGA1], Prop.2.1, p.30) (45.18): *Let $f : X \to Y$ be a morphism of schemes. Then the set of points of X where f is smooth is open.*

Lemma ([SGA1], Theorem 2.1, p.31) (45.19): *Let $f : X \to Y$ be a morphism locally of finite type. Let $x \in X$ and let $y = f(x)$. In order for f to be smooth at x, it is necessary and sufficient that f be flat at x and that $f^{-1}(y)$ be smooth over $\kappa(y)$ at x.*

§46 The Scheme Theoretic Argument

Let X_0, X_1, X_2, X_3, X_4 be indeterminates. For every triple i, j, k with $0 \le i \le j \le k \le 4$, let a_{ijk} be another indeterminate. We then consider the projective schemes

$$\mathbf{P}^4 = Proj(\mathbf{Z}[X_0, X_1, X_2, X_3, X_4])$$
$$\mathbf{P}^{34} = Proj(\mathbf{Z}[a_{ijk} \mid 0 \le i \le j \le k \le 4])$$

and their product $\mathbf{P}^{34} \times \mathbf{P}^4$. The bihomogeneous form

$$\mathcal{F} = \sum_{0 \le i \le j \le k \le 4} a_{ijk} X_i X_j X_k,$$

is a section of the locally free sheaf $\mathcal{O}(1) \boxtimes \mathcal{O}(3)$ on $\mathbf{P}^{34} \times \mathbf{P}^4$, where $\boxtimes$ denotes external tensor product. Accordingly the equation

$$\mathcal{F} = 0$$

defines a closed subscheme Λ of $\mathbf{P}^{34} \times \mathbf{P}^4$. The projection of Λ onto the first factor makes Λ into a $\mathbf{P}^{34}$-scheme which we call the **universal cubic threefold**. We will obtain other $\mathbf{P}^{34}$-schemes from systems of bihomogeneous equations in the same way without explicit mention of the relevant locally free sheaves.

By taking the second partials of $\mathcal{F}$ with respect to the variables X_0, X_1, X_2, X_3, X_4, we obtain a 5×5 symmetric matrix which we denote $M(\mathcal{F})$. Denote by $\mathcal{H}_k(\Lambda)$ the $\mathbf{P}^{34}$-scheme defined by the vanishing of all of the $k \times k$ minors of $M(\mathcal{F})$. We will call the $\mathbf{P}^{34}$-scheme $\mathcal{H}_5(\Lambda)$ the **Hessian** of Λ. We will call the $\mathbf{P}^{34}$-scheme $\mathcal{H}_4(\Lambda)$ the rank 3 locus of the Hessian of Λ.

Lemma (46.1): $\mathbf{P}^{34}$ *is geometrically unibranch at every point.*

Proof: This follows from the fact that the ring $\mathbf{Z}$ of integers is normal and entire and therefore so are all the local rings of any polynomial ring over $\mathbf{Z}$.

Lemma (46.2): *Every fibre of* $\mathcal{H}_4(\Lambda)$ *has dimension* ≥ 1.

Proof:

Corollary (46.3): *The set U of all points x of $\mathbf{P}^{34}$ such that the fibre of $\mathcal{H}_4(\Lambda)$ at x is a smooth irreducible curve is open and the projection p_1 is flat over that open subset.*

Proof: Since we are dealing with Noetherian schemes, the set $Flat(P_1)$ of all points x of $\mathbf{P}^{34}$ such that p_1 is flat at all points of $p_1^{-1}(x)$ is open. To see that $Flat(p_1)$ is nonempty, it suffices to show that it contains the point of $\mathbf{P}^{34}$ corresponding to Klein's cubic threefold Λ_0. Since $\mathbf{H}_4(\Lambda_0)$ is a symmetric determinantal variety of the right dimension, it is Cohen-Macaulay (cf. [Eis1], Theorem 18.18 and the remarks following it). In fact, denoting by $\mathbf{P}^{14}$ the projective space of 5×5 symmetric matrices, the locus Σ of matrices of rank ≤ 3 is a subvariety of codimension 3 whose natural desingularization is the variety $\widehat{\Sigma}$ of pairs (M, L) consisting of a matrix of rank ≤ 3 and L is a 2 plane annihilated by M. The projection of $\widehat{\Sigma}$ onto its first coordinate is a desingularization since it is an isomorphism over the points of Σ with rank exactly 3 and since $\widehat{\Sigma}$ is easily shown to be a projectivized vector bundle on the Grassmannian of 2 planes in 5 space. A 5×5 symmetric matrix whose entries are linear forms in 5 variables determines a linear mapping of $\mathbf{P}^4$ into $\mathbf{P}^{14}$ and therefore a $\mathbf{P}^4$ in $\mathbf{P}^{14}$. The intersection of that $\mathbf{P}^4$ with the locus Ω of matrices of rank < 3 is empty generically and in the case of Klein's cubic Λ_0. So we can identify $\mathcal{H}_4(\Lambda_0)$ with a curve in $\widehat{\Sigma}$. Since $\mathbf{P}^4$ in $\mathbf{P}^{14}$ is the intersection of 10 hyperplanes, $\mathcal{H}_4(\Lambda_0)$ in $\widehat{\Sigma}$ is the complete intersection of 10 hypersurfaces and is therefore Cohen-Macaulay. The explicit computation of the tangent space to $\mathcal{H}_0(\Lambda_0)$ at the coordinate points of $\mathbf{P}^4$ shows that it is reduced at such points. Since $\mathcal{H}(\Lambda_0)$ is irreducible and Cohen-Macaulay, it follows (cf. Exercise 18.9 of [Eis1]) that it is reduced everywhere. Furthermore, we know that $\mathcal{H}_4(\Lambda_0)$ is a smooth curve (cf. [A16], Theorem 38.9[24] By Lemma 45.20, p_1 is smooth at all of the points of $\mathcal{H}_4(\Lambda_0)$. By Lemma 45.18, the

---

[24] In [A16] and earlier publications, we didn't worry about embedded components and other scheme theoretic subtleties. This is the first time we have explicitly addressed such questions.

set of points where p_1 is smooth is open and by Lemma 45.19, p_1 is flat at all such points. Thus U is nonempty and certainly contained in $Flat(p_1)$. By Lemma 45.8 and Corollary 46.3, the locus of points where the fibre is smooth and geometrically irreducible is open. So U is itself open and p_1 is flat over U.

Lemma (46.4): *Denote by U the subset of $\mathbf{P}^{34}$ consisting of all points x of $\mathbf{P}^{34}$ such that the fibre of $\mathcal{H}_4(\Lambda)$ over x is a smooth irreducible curve of genus 26. Then U is a nonempty Zariski open set.*

Proof: By Lemma 45.9, the Euler characteristic of the fibre is locally constant, hence constant since U is irreducible. The fact that the fibres are all geometrically smooth and irreducible implies that, in the notation of Lemma 45.9, the function h_0 is locally constant on U, hence constant. Therefore, the function h_1, which gives the genera of the fibres of $\mathcal{H}_4(\Lambda)$, is also constant on U.

We will denote by $\mathcal{C}$ the preimage of U in $\mathcal{H}_4(\Lambda)$.

Lemma (46.5): *The relative Picard scheme $\mathcal{P}ic^0_{C/U}$ is an abelian scheme all of whose fibres are abelian varieties of dimension 26.*

Proof: This follows from Lemma 45.11.

Since all of these constructions are functorial, we have

Lemma (46.6): *The group GL_5 acts in a natural way on $\mathbf{P}^{34}$, on $\mathbf{P}^4$, on Λ, etc. Furthermore, if $\mathcal{A}$ is an abelian U-scheme equivariant for the action of GL_5 and if T is an endomorphism of $\mathcal{A}$, then T commutes with the action of GL_5.*

Proof: The only assertion that is not completely obvious is the last one. It, however, follows from the fact that any rational mapping from an affine algebraic group into an abelian variety is constant.[25] Alternatively, one can argue from the fact that the endomorphism ring of an abelian variety is discrete.

We have a correspondence Ξ of bidegree $(4,4)$ on $\mathcal{C}$ which we will define later. It induces an endomorphism of $\mathcal{P}ic^0_{C/U}$ which we denote T_Ξ.

Lemma (46.7): *Let S be a scheme, let $\mathcal{A}$ be an abelian S-scheme all of whose fibres have the same dimension and let T be an endomorphism of $\mathcal{A}$. Denote by $\mathcal{A}'$ the kernel of T and by $\mathcal{A}''$ the image of T. Then all of the fibres of $\mathcal{A}'$ (resp. $\mathcal{A}''$) have the same dimension.*

Proof: At every point s of S, we have

$$dim_s\mathcal{A} = dim_s\mathcal{A}' + dim_s\mathcal{A}''.$$

---

[25] For this argument, it is necessary to point out that the abelian scheme $\mathcal{A}$ is induced from an abelian scheme on the moduli space of cubic threefolds, at least over an open subset of U. Therefore, the restriction of $\mathcal{A}$ to a generic orbit X of GL_5 is a trivial family $X \times A$. Choosing a point a of A, one gets a morphism $GL_5 \to A$ as the composition

$$GL_5 \to X \to X \times a \to X \to A \xrightarrow{T} A.$$

By Lemma 45.1, the two dimensions on the right hand side are upper semicontinuous functions of s. Since their sum is, by hypothesis, constant, both of these semicontinuous functions must be constant.

Theorem (46.8):   *The abelian U-scheme $\mathcal{P}ic^0_{C/U}$ is the sum of abelian U-schemes $\mathcal{A}_{10}$ and $\mathcal{A}_{16}$ of dimensions 10 and 16 respectively. Each of these abelian U-schemes is invariant under the endomorphism T_Ξ. On $\mathcal{A}_{10}$, the endomorphism T_Ξ induces the endomorphism $x \mapsto 2x$, while on $\mathcal{A}_{16}$ it induces the endomorphism $x \mapsto -x$.*

Proof: Let us write $\mathcal{A}$ instead of $\mathcal{P}ic^0_{C/U}$. Applying the preceding lemma to the endomorphism $T_\Xi - 2_\mathcal{A}$, we conclude that all of the fibres of its kernel (respectively its image) have the same dimension. The same applies to the endomorphism $T_\Xi + 1_\mathcal{A}$. Let u denote the point of U determined by Klein's cubic in characteristic 0. On the fibre $\mathcal{A}_x$, the endomorphism T_Ξ induces the Hecke correspondence T_3, whose eigenvalues on the tangent space of the Jacobian variety of $X(11)$ are -1, with multiplicity 16, and 2, with multiplicity 10. Accordingly, the fibre over u of the kernel (resp. the image) of $T_\Xi - 2_\mathcal{A}$ has dimension 16 (resp. 10) and the fibre of the kernel (resp. the image) of $T_\Xi + 1_\mathcal{A}$ has dimension 10 (resp. 16). Since all the fibres of the image (resp. the kernel) of $T_\Xi - 2_\mathcal{A}$ have the same dimension, we conclude that the image of $T_\Xi - 2_\mathcal{A}$ is an abelian U-scheme $\mathcal{A}_{16}$ of fibre dimension 16 and the connected component of the identity of the kernel of $T_\Xi - 2_\mathcal{A}$ is an abelian U-scheme $\mathcal{A}_{10}$. Similarly, the image of the endomorphism $T_\Xi + 1_\mathcal{A}$ is an abelian U-scheme $\mathcal{B}_{10}$ of fibre dimension 10 and the connected component of the identify of the kernel of $T_\Xi + 1_\mathcal{A}$ is an abelian U-scheme $\mathcal{B}_{16}$ of fibre dimension 16. Clearly, the endomorphism $T_\Xi - 2_\mathcal{A}$ maps $\mathcal{B}_{16}$ onto itself and the endomorphism $T_\Xi + 1_\mathcal{A}$ maps $\mathcal{A}_{10}$ onto itself. Therefore, we have $\mathcal{A}_{10} = \mathcal{B}_{10}$ and $\mathcal{A}_{16} = \mathcal{B}_{16}$. Finally, since the difference of the endomorphisms $T_\Xi - 2_\mathcal{A}$ and $T_{\Xi+\mathcal{A}}$ is $3_\mathcal{A}$, which maps $\mathcal{A}$, $\mathcal{A}_{10}$ and $\mathcal{A}_{16}$ onto themselves, we conclude that

$$\mathcal{A} = \mathcal{A}_{10} + \mathcal{A}_{16}.$$

Lemma (46.9):   *The abelian schemes $\mathcal{A}_{10}$ and $\mathcal{A}_{16}$ are invariant under the action of GL_5.*

Proof: This follows from Lemma 46.6 with $T = T_\Xi - 2_\mathcal{A}$ and $T = T_\Xi + 1_\mathcal{A}$.

Corollary (46.10):   *The abelian scheme $\mathcal{A}_{10}$ is simple.*

Proof: If not, there is an endomorphism T of $\mathcal{A}_{10}$ mapping $\mathcal{A}_{10}$ onto an abelian subscheme $\mathcal{A}'$ of $\mathcal{A}_{10}$. By Lemma 46.8, the endomorphism T must commute with the action of GL_5. Therefore, the fibre over Klein's cubic u in characteristic 0 is an abelian variety on which $PSL_2(\mathbf{F}_{11})$ acts. The action of $PSL_2(\mathbf{F}_{11})$ on the holomorphic 1-forms on $C_u = X(11)$ is well known to decompose into irreducible representations of degrees 5, 10 and 11. From the equivariant decomposition of the space into the 10 dimensional tangent space to $\mathcal{A}_{10u}$ and the 16 dimensional tangent space of $\mathcal{A}_{16u}$, we know that the action of $PSL_2(\mathbf{F}_{11})$ on the tangent space to $\mathcal{A}_{10u}$ is irreducible. Therefore, there can be no $PSL_2(\mathbf{F}_{11})$-invariant abelian subvariety of $\mathcal{A}_{10u}$. Therefore the dimension of the fibre $\mathcal{A}'_u$ is either 0 or 10. Since all the fibres of $\mathcal{A}'$ must have the same dimension, we conclude that either $\mathcal{A}' = 0$ or $\mathcal{A}' = \mathcal{A}_{10}$. In particular, $\mathcal{A}_{10}$ is simple.

Corollary (46.11): *The abelian scheme $\mathcal{A}_{16}$ is either simple or else is the sum of an abelian scheme $\mathcal{A}_5$ of dimension 5 and and abelian scheme $\mathcal{A}_{11}$ of dimension 11.*

Proof: This is quite similar to the proof of the preceding lemma. The only essential difference is that instead of having the irreducible representation of $PSL_2(\mathbf{F}_{11})$ on the tangent space to the identity of the 10 dimensional abelian variety $\mathcal{A}_{10_u}$, one has the reducible representation on the tangent space to the identity of the 16 dimensional abelian variety $\mathcal{A}_{10_u}$. That representation decomposes into irreducible representations of dimensions 5 and 11 and from this the lemma follows.

Theorem (46.12): *The abelian U-schemes $\mathcal{A}_{10}$ and $\mathcal{A}_{16}$ are simple.*

Remark (46.13): This theorem asserts that $\mathcal{A}_{10}$ and $\mathcal{A}_{16}$ are simple as abelian schemes defined over U. This is equivalent to the assertion that the fibres of $\mathcal{A}_{10}$ and $\mathcal{A}_{16}$ over the generic point of U are simple over the function field of U, but not necessarily over the algebraic closure of the function field. In particular, for all we know at this stage, every fibre of $\mathcal{A}_{10}$ of $\mathcal{A}_{16}$ could be isogenous to the product of elliptic curves. We have merely asserted that no such decomposition is rational over U, nor can it be given by a geometric construction in the sense of [A6]. This distinction was pointed out to the author by Ofer Gabber. Torsten Ekedahl, who was kind enough to read portions of this article, pointed out how one can modify the proof of this theorem so as to draw the stronger conclusion that the abelian schemes $\mathcal{A}_{10}$ and $\mathcal{A}_{16}$ are in fact absolutely simple. We present his argument at the end of this section.

Proof of Theorem 46.12: In order to show that $\mathcal{A}_{16}$ is simple, we just have to prove that there does not exist an endomorphism of $\mathcal{P}ic^0_{C/U}$ whose image is an abelian scheme of fibre dimension 5. This will be accomplished by a sequence of lemmas. However, we note first that the abelian variety $\mathcal{A}_u$ decomposes into the sum of 3 abelian varieties of dimensions 5, 10 and 11 invariant under the group $PSL_2(\mathbf{F}_{11})$. Furthermore, since the component of dimension 5 is invariant under this group, it is isomorphic to the product of 5 copies of the elliptic curve E with complex multiplication by the ring of integers of the imaginary quadratic field $\mathbf{Q}(\sqrt{-11})$. Since this is the unique $PSL_2(\mathbf{F}_{11})$ invariant abelian variety of dimension 5 contained in $\mathcal{A}_u$, this abelian variety must be the fibre over u of the image of T.

$\qquad$For the arguments that follow, we will need to tensor the entire situation with $Spec(\mathbf{F}_3)$. If X is any object that we have been discussing over $Spec(\mathbf{Z})$, we will denote by $\underline{X}$ the resulting object over $Spec(\mathbf{F}_3)$.

$\qquad$Denote by Π the projective 4-plane in $\underline{\mathbf{P}^{34}}$ associated to the linear space $\mathcal{R}$ spanned by the cubic forms x_i^3 for $0 \le i \le 4$. Over a perfect field of characteristic 3, it is the same to describe Π as associated to forms which are cubes of linear forms. Let $\mathcal{K}$ denote the projective space of dimension 29 over $Spec(\mathbf{F}_3)$ consisting of projective 5-planes in $\underline{\mathbf{P}^{34}}$ containing Π. Then both Π and $\mathcal{K}$ are invariant under PGL_5. There is a rational mapping $\phi : \underline{\mathbf{P}^{34}} \to \mathcal{K}$ which associates to a point p not lying in Π the projective 5-plane spanned by Π and p.

Lemma (46.14): *The rational mapping ϕ is defined at all of the closed points of $\underline{U}$.*

Proof: It is obvious that $\underline{U}$ and Π are disjoint, since the Hessian of a cubic in Π vanishes identically.

We will denote by V the image of U under ϕ. We will denote by ϕ' the restriction of ϕ to U.

Lemma (46.15): *V is an open subscheme of $\mathcal{K}$. The fibres of ϕ' are 5 dimensional affine spaces over $\mathbf{F}_3$.*

Proof: The linear space $\mathcal{R}$ is invariant under GL_5, hence the semidirect product $GL_5 \cdot \mathcal{R}$ acts as a group of affine transformations on the 35 dimensional space S_3 of all cubic forms. We then have a GL_5 equivariant mapping

$$\tilde{\phi} : S_3 \to S_3/\mathcal{R}$$

whose fibres are affine spaces of dimension 5. Modulo the action of G_m, the mapping $\tilde{\phi}$ gives rise to the rational map

$$\phi : \mathbf{P}(S_3) = \underline{\mathbf{P}^{34}} \to \mathbf{P}(S_3/\mathcal{R}) = \mathcal{K}$$

with the same fibres as $\tilde{\phi}$, hence affine spaces of dimension 5.

Consider the product of V with $\underline{\mathbf{P}^4}$. Denote by $\mathcal{F}'$ the bihomogeneous form

$$\mathcal{F} = \sum a_{ijk} X_i X_j X_k,$$

where the summation runs over all triples i, j, k such that $0 \leq i \leq j \leq k \leq 4$ and $i \neq k$. Let M' denote the matrix of second partial derivatives of $\mathcal{F}'$ with respect to the variables $X_0, \ldots, X_4$. Denote by $\mathcal{C}'$ the locus in $V \times_{\mathbf{F}_3} \underline{\mathbf{P}^4}$ defined by the vanishing of the 4×4 minors of M'.

We leave the proof of the following easy lemma to the reader.

Lemma (46.16): *The fibre product of $\underline{U}$ and $\mathcal{C}'$ over V is $\underline{\mathcal{C}}$.*

Corollary (46.17): *The fibre product of $\underline{U}$ and the abelian scheme $\mathcal{A}' = \mathcal{P}ic^0_{\mathcal{C}'/V}$ over V is*

$$\underline{\mathcal{P}ic^0_{\mathcal{C}/U}} = \mathcal{P}ic^0_{\underline{\mathcal{C}}/\underline{U}}.$$

Proof: Obvious, given the above equality. The equality itself is a special case of the fact that Pic^0 commutes with base change, i.e. Corollary 45.15.

Thus, we have shown that the $\underline{U}$-scheme $\underline{\mathcal{C}}$ and its associated abelian scheme $\mathcal{P}ic^0_{\mathcal{C}/U}$ descend to V-schemes. Now, suppose $\mathcal{T}$ is an endomorphism of the abelian $\underline{U}$-scheme $\mathcal{A} = \mathcal{P}ic^0_{\mathcal{C}/U}$ such that the image of $\mathcal{T}$ is an abelian scheme of dimension 5 over U. Reducing modulo 3, we obtain from $\mathcal{T}$ an endomorphism $\underline{\mathcal{T}}$ of $\underline{\mathcal{A}}$ whose image has fibre dimension 5. Furthermore, $\underline{\mathcal{T}}$ must commute with $\underline{PGL_5}$.

Lemma (46.18): *The endomorphism $\underline{\mathcal{T}}$ descends to $\mathcal{A}'$. In other words, there is an endomorphism $\mathcal{T}'$ of $\mathcal{A}'$ such that the endomorphism induced by $\mathcal{T}'$ on the fibre product $\underline{\mathcal{A}} = \mathcal{A}' \times_{\mathbf{F}_3} \underline{U}$ is $\underline{\mathcal{T}}$.*

Proof: The argument is similar to that of Lemma 46.6. One can argue that since $\underline{\mathcal{A}}$ descends to $\mathcal{A}'$, the restriction of $\underline{\mathcal{A}}$ to one of the affine spaces that are the fibres of ϕ is a trivial family. If the endomorphism $\underline{\mathcal{T}}$ doesn't descend, then one gets a morphism from affine space into an abelian variety, which is therefore constant. Alternatively, one can argue that one gets a mapping from affine space into the endomorphism ring of the abelian scheme, but the endomorphism ring is discrete, hence the mapping is constant.

The following lemma is straightforward and is left to the reader.

Lemma (46.19): *The group $\underline{PGL_5}$ acts on V, C' and A' in such a way that the induced actions on $\underline{U}$, $\underline{C}$ and $\underline{A}$ are the actions we have already defined. Furthermore, the action of $\underline{GL_5}$ on A' commutes with T'.*

Denote by A_5' the image of T'. Denote by v the point of V determined by Klein's cubic, viewed as having coefficients in $\mathbf{F}_3$.

Lemma (46.20): *The Mathieu group M_{11} is a subgroup of $\underline{PGL_5}$ fixing v.*

Proof: This is the main result of the author's paper [A13].

We can now complete the proof of theorem 46.12. The fibre A_{5v}' of A_5' over v is the reduction modulo 3 of the fibre of A_5. Since the latter is the product of 5 copies of the elliptic curve E, the fibre A_{5v}' must be the product of 5 copies of the elliptic curve $\underline{E}$ obtained by reducing E modulo 3. Therefore the action of the group M_{11} on A_{5v}' gives rise to a nontrivial homomorphism of M_{11} into $GL_5(End(\underline{E}))$. Since the prime 3 splits in $\mathbf{Q}(\sqrt{-11})$, the endomorphism ring of $\underline{E}$ is the ring of integers of $\mathbf{Q}(\sqrt{-11})$. Therefore, the action of M_{11} on A_{5v}' gives rise to a nontrivial homomorphism

$$M_{11} \to GL_5(End(\underline{E})) \to GL_5(\mathbf{Q}(\sqrt{-11})) \to GL_5(\mathbf{C}).$$

But it is well known that the Mathieu group M_{11} has no nontrivial complex representation of degree 5. This contradicts the hypothesis that A_{16} is not simple. Therefore, the abelian U-scheme A_{16} is simple and the theorem is proved.

As promised earlier in this section, we now present Torsten Ekedahl's proof that the abelian schemes A_{10} and A_{16} are in fact absolutely simple. First we need a lemma.

Lemma (46.21): *Let B be an abelian scheme over U. If U is not absolutely simple, then B decomposes over an étale cover of U.*

Proof: Denote by $U_{\acute{e}t}$ the big etale site of U and by Ens the category of sets. Denote by $F : U_{\acute{e}t} \to Ens$ the functor which associates to each U-scheme V the set of endomorphisms of the abelian scheme B. Then F is a sheaf on $U_{\acute{e}t}$. By a Hilbert scheme arguemnt, F is representable by a U-scheme $\mathcal{E}$ locally of finite type. If K is a finite extension of the function field of U and if T is an endomorphism of $B \times_U Spec(K)$ over $Spec(K)$, then by the universal property of $\mathcal{E}$, there is a U-morphism of $Spec(K)$ to $\mathcal{E}$ which induces this endomorphism. Let Q be the point of $\mathcal{E}$ to which $Spec(K)$ is mapped by this U-morphism. Since the mapping is a U-morphism, the point Q lies over the generic point of U. Denote by D the closure of Q in $\mathcal{E}$. Then D is of finite type over U. Furthermore, $\mathcal{E}$ satisfies the valuative criterion for properness but it is not necessarily proper since it is not of finite type. However, D will then also satisfy the valuative criterion for properness and is of finite type, so D is proper. Since the structure morphism of the U-scheme D maps the point Q of D to the generic point of U, it follows that D is mapped surjectively to U. Since the endomorphism T is induced from an endomorphism of $\mathcal{B} \times_U D$, it will be enough to show that D is étale over U. Let d be any point of D lying over a point u of U. Then we have a natural homomorphism $\mathcal{O}_u \to \mathcal{O}_d$ of the local rings $\mathcal{O}_u$ of U at u into the local ring $\mathcal{O}_d$ of D at d. Denote by H_u, H_d the strict henselizations of $\mathcal{O}_u$, $\mathcal{O}_d$ respectively. Then we have an induced homomorphism from $H_u \to H_d$ and it is enough to prove that this homomorphism is an isomorphism. That it is injective follows from the fact that U is normal. That it is surjective is shown by a deformation theory argument, i.e. deformation of an endomorphism. The situation is as follows. Let $R \to S$ is a surjective homomorphism of Artinian local algebras and that we have morphisms $Spec(R) \to U$ and $Spec(S) \to D$ which make the obvious diagram commute. Then what must be shown is that any lifting of the morphism $Spec(R) \to U$ to a morphism $Spec(R) \to D$, such that the diagram still commutes, is unique. One can, by induction on the length of R minus the length of S, assume that the kernel of $R \to S$ is a vector space of dimension 1 over the residue field k of R. Furthermore, it is convenient to identify an endomorphism with its graph. So we are reduced to deforming an abelian subvariety of a fixed abelian variety. The space of lifts from S to R can be identified with the space of sections normal bundle of the abelian subvariety, i.e. of the graph. Since the normal bundle is trivial, one gets a vector space over k whose dimension is the codimension of the abelian subvariety. However, one is not merely deforming the abelian subvariety into other subvarieties; one is deforming it into other *abelian* subvarieties, in particular into subvarieties containing the identity element of the abelian variety. This condition forces the normal section to vanish at the identity element of the abelian subvariety and, since the normal bundle is trivial, therefore to vanish identically. This proves the uniqueness of the lifting and the lemma follows.

Theorem (46.22): *The abelian schemes $\mathcal{A}_{10}$ and $\mathcal{A}_{16}$ are absolutely simple.*

Proof: Let i denote either 10 or 16. Suppose $\mathcal{A}_i$ is not absolutely simple. Then there is an etale cover D of U such that $\mathcal{A}_i$ decomposes over D. We can assume that D is Galois over U. Accordingly, the Galois group Γ of D over U acts on the endomorphism ring of $\mathcal{A} \times_U D$. Every element of SL_5 induces a morphism of U to itself which lifts to a morphism of D to itself. The liftings so obtained form a group $\widehat{SL_5}$ which maps onto SL_5 with kernel Γ. Since SL_5 is simply connected and Γ is finite, the group $\widehat{SL_5}$ is actually isomorphic to the product of SL_5 and Γ. Let d_0 denote a point of D lying over the point u_0 of U corresponding to Klein's cubic threefold. Since the preimage of $PSL_2(\mathbf{F}_{11})$ in $SL_5(\mathbf{C})$ is $PSL_2(\mathbf{F}_{11}) \times \{\pm 1\}$, it follows that the group $PSL_2(\mathbf{F}_{11}) \times \{\pm 1\} \times \Gamma$ then acts transitively on the fibre of D over u_0. It follows that the isotropy group of d_0 in $\widehat{SL_5}$ contains a group isomorphic to $PSL_2(\mathbf{F}_{11})$. As in Lemma 46.6, the endomorphism T must commute with the action of $\widehat{SL_5}$. The endomorphism induced by T on the abelian variety which is the fibre of $\mathcal{A}_i$ over d_0 must therefore commute with the action of $PSL_2(\mathbf{F}_{11})$, which may be identified with the fibre and action over u_0. In particular, there cannot be a nonscalar endomorphism if the representation of $PSL_2(\mathbf{F}_{11})$ on the tangent space of the abelian variety is irreducible. This already proves that $\mathcal{A}_{10}$ is a simple abelian scheme. It also implies that if $\mathcal{A}_{16}$ does decompose, it can only decompose into abelian schemes of fibre dimensions 5 and 11. Having disposed of the case $i = 10$, we confine ourselves now to the case $i = 16$. In the case of the Klein cubic threefold, the fibre does in fact decompose in this manner, with the 5 dimensional part being the product of 5 copies of an elliptic curve E_5 with complex multiplication by $\mathbf{Q}(\sqrt{-11})$ and the 11 dimensional part being a product of 11 copies of an elliptic curve E_{11} which does not have complex multiplication. Since both the 5 and the 11 dimensional representations of $PSL_2(\mathbf{F}_{11})$ on the tangent spaces of these 5 dimensional abelian varieties are irreducible, it follows that the endomorphism ring of $\mathcal{A}_{16} \times_U D$ is contained in the commutant of $PSL_2(\mathbf{F}_{11})$ in the endomorphism ring of the 16 dimensional part of $Jac(X(11))$, i.e. in $M_5(End(E_5)) \oplus M_{11}(End(E_{11}))$. Hence the endomorphism ring of $\mathcal{A}_{16} \times_U D$ is contained in $\mathbf{Q} \oplus \mathbf{Q}(\sqrt{-11})$. Since the Galois group Γ acts as a group of automorphisms of the endomorphism algebra of $\mathcal{A}_{16} \times_U D$, we can assume that Γ is has order 1 or 2. If 1, we are done since we already know that $\mathcal{A}_{16}$ is simple over U, so we can assume that Γ has order 2. We can now proceed as in the proof that $\mathcal{A}_{16}$ is simple over U. Indeed, reducing the whole situation modulo 3, we get a 2-sheeted (hence tame at 3) cover $\underline{D}$ of $\underline{U}$ on which there is an endomorphism $\underline{T}$ with 5 dimensional image and commuting with the automorphisms of the abelian variety $\underline{\mathcal{A}}_{16}$. As before, we can further replace $\mathcal{U}$ by the variety V whose points correspond to the cubics of U modulo cubes of linear forms and $\underline{D}$, $\underline{\mathcal{A}}_{16}$, $\underline{T}$ will descend to a covering D' of V, an abelian scheme $\mathcal{A}'$ and an endomorphism T' of $\mathcal{A}'$, all invariant under SL_5. Again the Mathieu group M_{11} will act and will commute with the action of T' and we will get the same contradiction as before, since M_{11} has no 5 dimensional representation over $\mathbf{Q}(\sqrt{-11})$. Thus, $\mathcal{A}_{16}$ must be absolutely simple.

§47   A canonical form for quinary cubics

As we have already learned, Klein's cubic

$$v^2 w + w^2 x + x^2 y + y^2 z + z^2 v$$

is the unique cubic invariant for the odd part of the Weil representation of $SL_2(\mathbf{F}_{11})$ on the space V^-. We may regard this cubic as defining an equivariant linear mapping from V^- to the dual of $Sym^2(V^-)$. As such it may be identified with a 5×5 symmetric matrix whose entries are linear forms in v, w, x, y, z, namely

(47.0)
$$\begin{pmatrix} w & v & 0 & 0 & z \\ v & x & w & 0 & 0 \\ 0 & w & y & x & 0 \\ 0 & 0 & x & z & y \\ z & 0 & 0 & y & v \end{pmatrix}$$

which is, up to a factor of 2, the matrix of second partials of Klein's cubic. But according to the results of [A-R], the fundamental intertwining operator defines an equivariant isomorphism of $Sym^2(V^-)$ onto $\bigwedge^2 V^+$. Using this, we obtain an equivariant mapping from V^- to the dual of $\bigwedge^2 V^+$, and this we may identify with a 6×6 skew symmetric matrix whose entries are linear forms in v, w, x, y, z. It is not difficult to compute the matrix and to find that it is

(47.1)
$$
\begin{pmatrix}
0 & v & w & x & y & z \\
-v & 0 & 0 & z & -x & 0 \\
-w & 0 & 0 & 0 & v & -y \\
-x & -z & 0 & 0 & 0 & w \\
-y & x & -v & 0 & 0 & 0 \\
-z & 0 & y & -w & 0 & 0
\end{pmatrix}.
$$

Lemma (47.2): *The Pfaffian of the matrix (47.1) is Klein's cubic*

$$v^2 w + w^2 x + x^2 y + y^2 z + z^2 v.$$

Proof: In view of the equivariance with respect to $PSL_2(\mathbf{F}_{11})$, the Pfaffian of this skew symmetric matrix will have to be a cubic invariant for $PSL_2(\mathbf{F}_{11})$ acting on V^- and will therefore have to be a scalar multiple of Klein's cubic. In fact, a direct computation, which the reader can easily do by hand using the formula for the Pfaffian in the proof of Theorem 47.3 below, shows that the Pfaffian of (47.1) is equal to Klein's cubic.

Having obtained one "randomly selected" cubic as a Pfaffian, and believing in the principle, discussed in [A6] and [A18], that there is really nothing inherently atypical about being an invariant of a finite group, it is natural to hope that a generic cubic can also be written as a Pfaffian. This is indeed the case and the proof uses in an essential way the fact that it is so for Klein's cubic.

Theorem (47.3): *A generic quinary cubic can be represented in ∞^5 essentially different ways as the Pfaffian of a skew-symmetric 6×6 matrix whose entries are linear forms in 5 variables.*

Proof: Let the variables be x_1, x_2, x_3, x_4, x_5. Denote by $\mathcal{P}$ the space of all such matrices and by $\mathcal{K}$ the space of all cubic forms in 5 variables. Denote by Pf the mapping from $\mathcal{P}$ to $\mathcal{K}$ induced by the Pfaffian. In order to prove the theorem, we just have to show that the mapping Pf is generically surjective. For that, we only need to show that its differential is surjective at some point of Pf. Let A be a typical element of $\mathcal{P}$ and let its $(i, j)-th$ entry be denoted A_{ij}, where A_{ij} is a linear form

$$A_{ij} = a_{ij1} x_1 + a_{ij2} x_2 + a_{ij3} x_3 + a_{ij4} x_4 + a_{ij5} x_5$$

in x_1, x_2, x_3, x_4, x_5 such that $A_{ii} = 0$ for all i and $A_{ji} = -A_{ij}$ for all i, j. Then $Pf(A)$ is given by

$$
\begin{aligned}
Pf(A) = {} & A_{12} A_{34} A_{56} - A_{12} A_{35} A_{46} + A_{12} A_{36} A_{45} \\
& - A_{13} A_{24} A_{56} + A_{13} A_{25} A_{46} - A_{13} A_{26} A_{45} \\
& + A_{14} A_{23} A_{56} - A_{14} A_{25} A_{36} + A_{14} A_{26} A_{35} \\
& - A_{15} A_{23} A_{46} + A_{15} A_{24} A_{36} - A_{15} A_{26} A_{34} \\
& + A_{16} A_{23} A_{45} - A_{16} A_{24} A_{35} + A_{16} A_{25} A_{34}.
\end{aligned}
$$

The partials of $Pf(A)$ with respect to the 75 indeterminates are surprisingly pleasant to describe. We have

$$\frac{\partial Pf(A)}{\partial a_{ijk}} = (-1)^{i+j+1} x_k (A_{pq} A_{rs} + A_{pr} A_{sq} + A_{ps} A_{qr}),$$

where we may assume that $i < j$, $p < q < r < s$ and the sets $\{i, j\}$ and $\{p, q, r, s\}$ are disjoint. Thus, the differential of Pf maps the tangent space to $\mathcal{P}$ at A onto the space of all cubics lying in the ideal generated by quadratics of the form

(47.4) $$Q_{ij} = A_{pq} A_{rs} + A_{pr} A_{sq} + A_{ps} A_{qr},$$

where i, j, p, q, r, s are as above. In the particular case of the matrix

(47.5) $$\begin{pmatrix} 0 & v & w & x & y & z \\ -v & 0 & 0 & z & -x & 0 \\ -w & 0 & 0 & 0 & v & -y \\ -x & -z & 0 & 0 & 0 & w \\ -y & x & -v & 0 & 0 & 0 \\ -z & 0 & y & -w & 0 & 0 \end{pmatrix},$$

whose Pfaffian is Klein's cubic

$$v^2 w + w^2 x + z^2 y + y^2 z + z^2 v,$$

one sees by a painless hand computation that the quadratics (47.4) one obtains form a basis for all quadratics in the 5 variables v, w, x, y, z. Indeed, the quadratics Q_{ij} are given by

$$\begin{pmatrix} 0 & Q_{12} & Q_{13} & Q_{14} & Q_{15} & Q_{16} \\ Q_{21} & 0 & Q_{23} & Q_{24} & Q_{25} & Q_{26} \\ Q_{31} & Q_{32} & 0 & Q_{34} & Q_{35} & Q_{36} \\ Q_{41} & Q_{42} & Q_{43} & 0 & Q_{45} & Q_{46} \\ Q_{51} & Q_{52} & Q_{53} & Q_{54} & 0 & Q_{56} \\ Q_{61} & Q_{62} & Q_{63} & Q_{64} & Q_{65} & 0 \end{pmatrix} =$$

$$\begin{pmatrix} 0 & -vw & -wx & -xy & -yz & -zv \\ vw & 0 & -wy & -y^2 - zv & w^2 + xy & xv \\ wx & wy & 0 & -zx & -vw - z^2 & x^2 + yz \\ xy & y^2 + zv & zx & 0 & -vy & -v^2 - wx \\ yz & -w^2 - xy & vw + z^2 & vy & 0 & -wz \\ zv & -xv & -x^2 - yz & v^2 + wx & wz & 0 \end{pmatrix}$$

The verification that these quadratics form a basis for all quadratics in v, w, x, y, z can be made less tedious if one notices that if one applies the cyclic permutation (23456) of order 5 to the subscripts of Q_{ij} (using $Q_{ji} = -Q_{ij}$ if necessary), the resulting quadratic is the same as that obtained by the cyclic permutation $(vwxyz)$ of the variables.

The ideal these quadratics generate therefore contains all cubics and that proves that the differential of Pf is surjective in this case. As for the number of essentially different representations of a generic cubic, we argue as follows. We can identify a generic matrix A in $\mathcal{P}$ with a linear embedding of $\mathbf{C}^5$ into the 15 dimensional space of alternating forms on $\mathbf{C}^6$. Up to a scalar factor, this is the same as giving an embedding of $\mathbf{P}^4$ into the 14 dimensional projective space $\mathbf{P}(\bigwedge^2(\mathbf{C}^6))$. The image of $\mathbf{P}^4$ under this mapping determines an element of the Grassmannian $Gr(4, 14)$ of $\mathbf{P}^4$'s in $\mathbf{P}^{14}$, which is a variety of dimension $5 \times 10 = 50$. The group $PGL_6(\mathbf{C})$ has dimension 35 and acts on the Grassmannian, the orbit space being of dimension[26] $50 - 35 = 15$. On the other hand, the 25 dimensional group $GL_5(\mathbf{C})$ acts on the 35 dimensional space $S_3(\mathbf{C}^5)$ of all cubic forms in 5 variables and has an orbit space of dimension 10. We therefore have a generically surjective mapping from the 15 dimensional orbit space $PGL_6(\mathbf{C})\backslash Gr(4, 14)$ onto the 10 dimensional moduli space $GL_5(\mathbf{C})\backslash S_3(\mathbf{C}^5)$ of cubic threefolds. The generic fibre therefore has dimension 5 and we are done.

§48   Vector Bundles On Cubic Threefolds

Lemma (48.1):   *The rank of the skew-symmetric matrix (47.1) is not less than 4 unless $v = w = x = y = z = 0$. In particular, there is no point $[v, w, x, y, z]$ of $\mathbf{P}^4$ for which the matrix has rank less than 4.*

Proof: For $1 \leq i < j \leq 6$, denote by $\mu(i, j)$ the 4×4 submatrix formed by crossing out the i-th and j-th rows and columns. Then we have

$$det(\mu(2, 3)) = w^2 y^2$$
$$det(\mu(3, 4)) = x^2 z^2$$
$$det(\mu(4, 5)) = v^2 y^2$$
$$det(\mu(5, 6)) = w^2 z^2$$
$$det(\mu(6, 2)) = v^2 x^2$$

Therefore, if (47.1) has rank < 4, we must have

$$vw = wx = xy = yz = zv = 0,$$

which implies that four of the variables v, w, x, y, z are zero. But one can verify directly that in that case, with the remaining variable non-zero, the rank of (47.1)

---

[26]   It is necessary to verify that there is at least one point of $Gr(4, 14)$ whose isotropy group in $PGL_6(\mathbf{C})$ is finite. The matrix (43.1) associated to Klein's cubic threefold gives a particular point x of $Gr(4, 14)$ fixed by the group $PSL_2(\mathbf{F}_{11})$. Since the only algebraic subgroups of $PGL_6(\mathbf{C})$ containing $PSL_2(\mathbf{F}_{11})$ are $PSL_2(\mathbf{F}_{11})$ itself, $PSO_6(\mathbf{C})$ and $PGL_6(\mathbf{C})$ itself (cf. [A17], for example) it follows that the isotropy group G_x of x must be one of these 3 groups. Since G_x leaves the linearly embedded $\mathbf{P}^4$ associated to x invariant, the representation of G_x on $\bigwedge^2(\mathbf{C}^6)$ must be reducible, having an invariant subspace of dimension 5. This is not the case for either $PGL_6(\mathbf{C})$ or $PSO_6(\mathbf{C}^6)$, which proves that the isotropy group G_x is finite in this case.

is equal to 4, which is a contradiction. Thus the lemma is proved.

Theorem (48.2): *Let M be a sufficiently generic element of $\mathcal{P}$. Then M give rise to a rank 2 vector bundle on the cubic*

$$Pf(M) = 0.$$

Proof: This follows at once from Lemma 48.1. However, the following argument, also works and involves no computation. Let $M = M_x$ be a generic 6×6 symmetric matrix whose entries are linear forms in 5 variables $x_0, \ldots, x_4$. Then M determines a generic linear subvariety L of dimension 4 in the projective space $\mathbf{P}^{14}$ of alternating forms on $\mathbf{C}^6$. The subvariety of $\mathbf{P}^{14}$ consisting of alternating forms of rank < 4 is a Grassmannian $Gr(\mathbf{P}^3, \mathbf{P}^5)$ of dimension 8. Since $4 + 8 < 14$, the generic linear subvariety L does not meet the Grassmannian. Therefore, the rank < 4 locus of M is empty and M_x has rank 2 for all x in the cubic hypersurface $Pf(M_x) = 0$. The locus of pairs (x, y) in $\mathbf{P}^4 \times \mathbf{C}^6$ such that $M_x \cdot y = 0$ is then a rank 2 vector bundle over the cubic via projection onto the first factor.

Notation (48.3): *If $M = M_x$ is a skew-symmetric 6×6 matrix whose entries are linear forms in 5 variables $x_0, \ldots, x_4$, and if the rank of M is never less than 4 for nonzero x, then the vector bundle constructed in the proof of the preceding theorem will be denoted $E(M)$. The cubic threefold $Pf(M) = 0$ will be denoted $\Lambda(M)$.*

Remark (48.4): Klein's embedding of $X(11)$ as a curve of degree 50 in $\mathbf{P}^4$ maps the curve into the Klein cubic threefold (cf. [K2], [A11]). Therefore, the ∞^5 rank 2 vector bundles we have constructed on the cubic threefold give rise to ∞^5 rank 2 vector bundles on $X(11)$. It would be interesting to know more about them. Presently, we do not even know the dimension of the moduli space of invariant rank 2 vector bundles on $X(p)$ in general since its possible components are described by specifying actions of isotropy groups of points of $X(p)$ on the fibres of the unknown vector bundle. Once that is specified, one can in principle compute the dimension of a given component, assuming it is nonempty. However, we do not know which components are nonempty. This is so even in the case of line bundles. Indeed, without actually stumbling onto the line bundles α and ζ corresponding to the A-curve and z-curve respectively, we would not know how to prove that they both exist.

§49   The jumping lines of $E(M)$

It is well known that a cubic threefold is a union of ∞^2 lines, these being represented by the points of the Fano surface F of the cubic. If E is a rank 2 vector bundle on a cubic threefold Λ and if L is a line lying in Λ, then the restriction of E to L is the direct sum of two line bundles, say

$$E \mid L = \mathcal{O}_L(n_1) \oplus \mathcal{O}_L(n_2),$$

where $n_1 = n_1(L)$ and $n_2 = n_2(L)$ depend on L. The value of the unordered pair $n_1(L), n_2(L)$ is a constructible function on the Fano surface $F = F(\Lambda)$ of the cubic. In particular, there is a Zariski open subset $U(E)$ of F for which the corresponding lines L exhibit the generic behavior. The lines for which the pair is not the same as for the generic line are called **jumping lines**. The locus of jumping lines is an important invariant of a vector bundle and we propose to say something about that locus in the case of our vector bundles $E(M)$.

Definition (49.1): *Let M be a sufficiently generic point of $\mathcal{P}$. Let L be a line contained in the cubic threefold $\Lambda(M)$. We will say that L is of the **first kind** if the kernels of the matrices M_x with $x \in L$ have no nonzero vector in common. If L is not of the first kind. we say that L is of the **second kind**.*[27]

Lemma (49.2): *Let M be as in the preceding definition. If L is a line of the first kind in $\Lambda(M)$, then the restriction of $E(M)$ to L is isomorphic to $\mathcal{O}(-1) \oplus \mathcal{O}(-1)$. If L is of the second kind, then the restriction of $E(M)$ to L is isomorphic to $\mathcal{O}(-2) \oplus \mathcal{O}$.*

Proof: As x varies over the points of the projective line L, we obtain a pencil of skew-symmetric forms M_x, each determined only up to a nonzero scalar factor. Let β_0, β_1 be two forms in the pencil. Let s, t be indeterminates. Then the general member of the pencil will be $s\beta_0 + t\beta_1$ and we denote this general member $\beta_{s,t}$. Since L lies in $\Lambda(M)$, $\beta_{s,t}$ has rank 4 for all $[s, t] \in \mathbf{P}^1$. We will denote by $e_1, \ldots, e_6$ a standard basis for $\mathbf{C}^6$.

Case 1: L **is of the first kind.** By assumption, there does not exist a nonzero vector common to the kernels of the matrices M_x for $x \in L$. Therefore, $ker(\beta_0) \cap ker(\beta_1) = (0)$. We can assume that β_0 is given by

$$\beta_0 = \begin{pmatrix} 0 & 1 & 0 & 0 & 0 & 0 \\ -1 & 0 & 0 & 0 & 0 & 0 \\ 0 & 0 & 0 & 1 & 0 & 0 \\ 0 & 0 & -1 & 0 & 0 & 0 \\ 0 & 0 & 0 & 0 & 0 & 0 \\ 0 & 0 & 0 & 0 & 0 & 0 \end{pmatrix}$$

We know that the kernels of β_0 and β_1 are disjoint but we have to consider whether the kernel of one of the matrices is isotropic for the other. If this is not the case, we can assume that the kernel of β_1 is spanned by e_1, e_2. The matrix β_1 is then of the form

$$\begin{pmatrix} 0 & 0 & 0 & 0 & 0 & 0 \\ 0 & 0 & 0 & 0 & 0 & 0 \\ 0 & 0 & 0 & a & b & c \\ 0 & 0 & -a & 0 & d & e \\ 0 & 0 & -b & -d & 0 & f \\ 0 & 0 & -c & -e & -f & 0 \end{pmatrix}$$

and the condition that β_1 have rank 4 is equivalent to the condition $af - be + cd \neq 0$.

---

[27] We should emphasize that our separation of lines into two kinds is unrelated to that of [C-G]. Theirs is based on the normal bundle of the line in the cubic threefold, i.e. whether it is isomorphic to $\mathcal{O} \oplus \mathcal{O}$ or to $\mathcal{O}(-1) \oplus \mathcal{O}(1)$.

The Pfaffian of the general member of the pencil $\beta_{s,t} = s\beta_0 + t\beta_1$ is then

$$st(fs + (af - be + cd)t),$$

which, since $af - be + cd \neq 0$, is nonzero for generic s, t. But that contradicts the assertion that the pencil consists of matrices of rank 4.

Therefore, each of the matrices β_0, β_1 has a kernel isotropic for the other. Accordingly, we can assume that the kernel of β_1 is spanned by e_1 and e_3 and that β_1 is of the form

$$\beta_1 = \begin{pmatrix} 0 & 0 & 0 & 0 & 0 & 0 \\ 0 & 0 & 0 & a & b & c \\ 0 & 0 & 0 & 0 & 0 & 0 \\ 0 & -a & 0 & 0 & d & e \\ 0 & -b & 0 & -d & f & 0 \\ 0 & -c & 0 & -e & -f & 0 \end{pmatrix}$$

with $af - be + cd \neq 0$. This form for β_1 follows from the assumption that β_1 has rank 4 and that e_1 and e_3 are in the kernel of β_1. But since we also know that the kernel of β_0 must be isotropic for β_1, we in fact have $f = 0$, so $be - cd \neq 0$.

Denoting by $x_0, \ldots, x_5$ homogeneous polynomials of the same degree δ in s, t and such that the column vector with entries $x_0, \ldots, x_5$ is in the kernel of $\beta_{s,t}$, we have

$$\begin{pmatrix} 0 \\ 0 \\ 0 \\ 0 \\ 0 \\ 0 \end{pmatrix} = \begin{pmatrix} 0 & s & 0 & 0 & 0 & 0 \\ -s & 0 & 0 & ta & tb & tc \\ 0 & 0 & 0 & s & 0 & 0 \\ 0 & -ta & -s & 0 & td & te \\ 0 & -tb & 0 & -td & 0 & 0 \\ 0 & -tc & 0 & -te & 0 & 0 \end{pmatrix} \begin{pmatrix} x_0 \\ x_1 \\ x_2 \\ x_3 \\ x_4 \\ x_5 \end{pmatrix} =$$

$$\begin{pmatrix} sx_1 \\ -sx_0 + t(ax_3 + bx_4 + cx_5) \\ sx_3 \\ -tax_1 - sx_2 + tdx_4 + tex_5 \\ -t(bx_1 + dx_3) \\ -t(cx_1 + ex_3) \end{pmatrix}$$

From the isolated first and third entries, sx_1 and sx_3, we conclude that $x_1 = x_3 = 0$. This substitution causes the last two entries to vanish, so the only remaining entries are $t(bx_4 + cx_5) - sx_0$ and $t(dx_4 + ex_5) - sx_2$. Therefore, since these vanish, t must divide x_0 and x_2 and we write $x_0 = ty_0$ and $x_2 = ty_2$. We then have the equality

$$(49.3) \qquad \begin{pmatrix} b & c \\ d & e \end{pmatrix} \begin{pmatrix} x_4 \\ x_5 \end{pmatrix} = s \begin{pmatrix} y_0 \\ y_2 \end{pmatrix}$$

Since $bd - ce \neq 0$, this matrix equation implies that s divides both x_4 and x_5, and we write $x_4 = sy_4$, $x_5 = sy_5$. Substituting these values in (49.4), we then have

$$(49.4) \qquad \begin{pmatrix} b & c \\ d & e \end{pmatrix} \begin{pmatrix} y_4 \\ y_5 \end{pmatrix} = \begin{pmatrix} y_0 \\ y_2 \end{pmatrix}$$

which gives us the value of y_0 and y_2. Therefore, we have

$$(49.5) \quad \begin{pmatrix} x_0 \\ x_1 \\ x_2 \\ x_4 \\ x_5 \end{pmatrix} = \begin{pmatrix} s(by_4 + cy_5) \\ 0 \\ s(dy_4 + ey_5) \\ 0 \\ sy_4 \\ sy_5 \end{pmatrix} = \begin{pmatrix} bs & cs \\ 0 & 0 \\ ds & es \\ 0 & 0 \\ s & 0 \\ 0 & s \end{pmatrix} \begin{pmatrix} y_4 \\ y_5 \end{pmatrix}$$

where y_4 and y_5 are both homogeneous of degree $\delta - 1$ and otherwise arbitrary. This proves that the restriction of $E(Q)$ to λ is isomorphic to $\mathcal{O}(-1) \oplus \mathcal{O}(-1)$.

Case 2: L is of the second kind. By definition, there is a nonzero vector common the kernels of β_0 and β_1. After a change of coordinates, the vector is e_6 and that the pencil is generated by the two forms

$$\beta_0 = \begin{pmatrix} 0 & 1 & 0 & 0 & 0 & 0 \\ -1 & 0 & 0 & 0 & 0 & 0 \\ 0 & 0 & 0 & 1 & 0 & 0 \\ 0 & 0 & -1 & 0 & 0 & 0 \\ 0 & 0 & 0 & 0 & 0 & 0 \\ 0 & 0 & 0 & 0 & 0 & 0 \end{pmatrix}$$

$$\beta_1 = \begin{pmatrix} 0 & a & b & 0 & c & 0 \\ -a & 0 & d & 0 & e & 0 \\ -b & -d & 0 & 0 & f & 0 \\ 0 & -e & 0 & 0 & 0 & 0 \\ -c & 0 & -f & 0 & 0 & 0 \\ 0 & 0 & 0 & 0 & 0 & 0 \end{pmatrix}$$

where $af - be + cd \neq 0$. This last condition guarantees that β_1 has rank 4. We then have

$$\beta_{s,t} = \begin{pmatrix} 0 & s + ta & tb & 0 & tc & 0 \\ -s - ta & 0 & td & 0 & te & 0 \\ -tb & -td & 0 & s & tf & 0 \\ 0 & 0 & -s & 0 & 0 & 0 \\ -tc & -te & -tf & 0 & 0 & 0 \\ 0 & 0 & 0 & 0 & 0 & 0 \end{pmatrix}.$$

We determine the kernel of $\beta_{s,t}$ acting on $\mathbf{C}[s,t]^6$. Let $x_0, x_1, x_2, x_3, x_4, x_5 \in \mathbf{C}[s,t]$ and suppose that the column vector with entries $x_0, \ldots, x_5$ is in the kernel of $\beta_{s,t}$. Then we have

$$\begin{pmatrix} 0 \\ 0 \\ 0 \\ 0 \\ 0 \\ 0 \end{pmatrix} = \begin{pmatrix} 0 & s + ta & tb & 0 & tc & 0 \\ -s - ta & 0 & td & 0 & te & 0 \\ -tb & -td & 0 & s & tf & 0 \\ 0 & 0 & -s & 0 & 0 & 0 \\ -tc & -te & -tf & 0 & 0 & 0 \\ 0 & 0 & 0 & 0 & 0 & 0 \end{pmatrix} \begin{pmatrix} x_0 \\ x_1 \\ x_2 \\ x_3 \\ x_4 \\ x_5 \end{pmatrix} =$$

$$\begin{pmatrix} sx_1 + t(ax_1 + bx_2 + cx_4) \\ -sx_0 + t(-ax_0 + dx_2 + ex_4) \\ sx_3 + t(-bx_0 - dx_1 + fx_4) \\ -sx_2 \\ -t(cx_0 + ex_1 - fx_2) \\ 0 \end{pmatrix}$$

Since the entry $-sx_2$ must vanish, we have $x_2 = 0$, Also, from the first three entries, we see that x_0, x_1 and x_3 must all be divisible by t. Writing $x_0 = ty_0$, $x_1 = ty_1$, $x_3 = ty_3$, we therefore have

$$\begin{pmatrix} sy_1 + aty_1 + cx_4 \\ -sy_0 - aty_0 - ex_4 \\ sy_3 - tby_0 + fx_4 \\ 0 \\ -t(cy_0 + ey_1) \\ 0 \end{pmatrix} = \begin{pmatrix} 0 \\ 0 \\ 0 \\ 0 \\ 0 \\ 0 \end{pmatrix}$$

Now, we cannot have $c = e = 0$, since otherwise, putting $s = -at$ leads to $\beta_{s,t}$ having rank < 4. Therefore the equation

$$cy_0 - ey_1 = 0,$$

from the fifth entry, implies that for some $u \in \mathbf{C}[s,t]$, we have

$$y_0 = eu, \quad y_1 = -cu.$$

By examining the third entry, we conclude that x_3 is divisible by t and write $x_3 = ty_3$. By examining the first two entries and using the fact that c, e are not both 0, we conclude that $x_4 = (s + ta)u$. This makes the first two entries vanish. The only remaining entry is the third, examination of which implies

(49.6) $s(y_3 - fu) = tu(af - be + cd).$

Since β_1 has rank 4, we have $af - be + cd \neq 0$. Therefore u must be divisible by s and we write $u = sv$. Therefore

$$y_3 = fsv + (af - be + cd)tv$$

and we conclude that

$$\begin{pmatrix} x_0 \\ x_1 \\ x_2 \\ x_3 \\ x_4 \\ x_5 \end{pmatrix} = \begin{pmatrix} stev \\ -stcv \\ 0 \\ tv(fs + (af - be + cd)t) \\ (s + ta)v \\ x_5 \end{pmatrix} = \begin{pmatrix} ste & 0 \\ -stc & 0 \\ 0 & 0 \\ t(fs + (af - be + cd)t) & 0 \\ (s + ta) & 0 \\ 0 & 1 \end{pmatrix} \begin{pmatrix} u \\ x_5 \end{pmatrix}.$$

If we require all the x_i to be homogeneous of some degree δ, the polnomials u and x_5 will be homogeneous of degrees $\delta - 2$ and δ, respectively, and otherwise arbitrary. That proves that the restriction to the line $\kappa(y)$ of the vector bundle $E(Q)$ on $\Lambda(Q)$ associated to Q is isomorphic to $\mathcal{O}(-2) \oplus \mathcal{O}$. This completes the proof of the lemma.

Lemma (49.7):   *Let M be a sufficiently generic point of $\mathcal{P}$. Then a generic line L lying in $\Lambda(M)$ is of the first kind.*

Proof: It is clear that being of the first kind is an open condition on L and having an L of the first kind is an open condition on M. So the point of this lemma is that the open sets in question is nonempty. Therefore, it will be enough to verify the lemma in the special case of (47.1). In this case, the line $v = w = x = 0$ lies in Klein's cubic threefold. If we set v, w, x equal to 0 in (47.1), the resulting pencil of matrices is spanned by

$$
\beta_0 = \begin{pmatrix} 0 & 0 & 0 & 0 & 1 & 0 \\ 0 & 0 & 0 & 0 & 0 & 0 \\ 0 & 0 & 0 & 0 & 0 & -1 \\ 0 & 0 & 0 & 0 & 0 & 0 \\ -1 & 0 & 0 & 0 & 0 & 0 \\ 0 & 0 & 1 & 0 & 0 & 0 \end{pmatrix}, \quad
\beta_1 = \begin{pmatrix} 0 & 0 & 0 & 0 & 0 & 1 \\ 0 & 0 & 0 & 1 & 0 & 0 \\ 0 & 0 & 0 & 0 & 0 & 0 \\ 0 & -1 & 0 & 0 & 0 & 0 \\ 0 & 0 & 0 & 0 & 0 & 0 \\ -1 & 0 & 0 & 0 & 0 & 0 \end{pmatrix}.
$$

Since the kernel of β_0 is spanned by e_2, e_4 and the kernel of β_1 is spanned by e_3, e_5, it is clear that this pencil is of the first kind and the lemma is proved.

Corollary (49.8):   *Let M be a sufficiently generic element of $\mathcal{P}$. Then the lines of the second kind are the jumping lines of the vector bundle $E(M)$.*

Proof: This follows immediately from the preceding two lemmas.
  The lines in $\mathbf{P}^4$ are the points of a variety, namely the Grassmannian variety $Gr(1, 4)$. The lines in a cubic threefold are the points of a surface F in $Gr(1, 4)$ called the **Fano surface** of the cubic threefold. When the cubic threefold is of the form $\Lambda(M)$ with M a sufficiently generic element of $\mathcal{P}$, we will write $F(M)$ to denote the Fano surface of $\Lambda(M)$.

Corollary (49.9):   *Let M be a sufficiently generic element of $\mathcal{P}$. Then the set of jumping lines of $E(M)$ is an algebraic subset of $F(M)$ of dimension ≤ 1.*

Proof: Since the jumping lines are the lines of the second kind, while the lines of the first kind form a nonempty open set, the statement is evident.

Lemma (49.10):   *Let M be a sufficiently generic element of $\mathcal{P}$. Let L be a line of the second kind in $\Lambda(M)$. Then there is one and only one point y of $\mathbf{P}^5$ such that $M_x y = 0$ for all $x \in L$.*

Proof: That there is such a point y is precisely the definition of "second kind". If y is not unique then let y' be another such point and let K be the 2-plane in $\mathbf{C}^6$ associated to the projective line joining y and y'. Then K lies in the kernel of M_x for all $x \in L$. Therefore, L determines a pencil of alternating forms on $\mathbf{C}^6/K$. Since a pencil of alternating forms on a 4-dimensional space must include at least one singular form, there will be a point $t \in L$ such that M_x has rank < 4, which contradicts Lemma 49.2. This proves that y must be unique.

Notation (49.11):   *If L is a line of the second kind, we will denote by $\varpi(L)$ the unique point $y \in \mathbf{P}^5$ associated to L as in the preceding lemma.*

§50   The associated quartic fourfold in $\mathbf{P}^5$

Let M be a sufficiently generic element of $\mathcal{P}$. For each point x of $\Lambda(M)$, the kernel of M_x is a 2-plane $E(M)_x$ in $\mathbf{C}^6$. Associated to each such 2-plane is a projective line $\mathbf{P}(E(M)_x)$ in the projective space $\mathbf{P}^5$ of lines in $\mathbf{C}^6$. Denote by $Q(M)$ the union of the lines $\mathbf{P}(E(M)_x)$ in $\mathbf{P}^5$. Denote by $\tilde{Q}(M)$ the "disjoint union" of these same projective lines. More precisely, $\tilde{Q}(M)$ is the subvariety of $\mathbf{P}^4 \times \mathbf{P}^5$ consisting of pairs (x, y) such that $y \in \mathbf{P}(E(M)_x)$. We can also describe $\tilde{Q}(M)$ as the $\mathbf{P}^1$ bundle $\mathbf{P}(E)$ associated to the vector bundle E and consisting of the lines in the fibres of E.

We denote by π_1, π_2 the restrictions to $\tilde{Q}(M)$ of the projections of $\mathbf{P}^4 \times \mathbf{P}^5$ onto its first and second coordinates respectively. If L is a line of the second kind in $\Lambda(M)$ then the mapping $x \mapsto \varpi(L)$, where ϖ is as defined at the end of the preceding section, defines a canonical section of π_1 over L. We will denote its image by $L^\#$ and we will refer to $L^\#$ as a line of the **third kind**.

Lemma (50.1):   *Let M be a sufficiently generic element of $\mathcal{P}$. Then $Q(M)$ is a quartic hypersurface in $\mathbf{P}^5$ and π_2 is a birational morphism of $\tilde{Q}(M)$ onto $Q(M)$. Denote by $\mathcal{L}(M)$ the union of all of the lines of the third kind in $\tilde{Q}(M)$. Then $\mathcal{L}(M)$ is a closed subvariety of codimension ≥ 2 in $\tilde{Q}(M)$ and π_2 maps the complement of $\mathcal{L}(M)$ in $\tilde{Q}(M)$ isomorphically onto its image.*

Proof: That π_2 is a morphism of $\tilde{Q}(M)$ with image $Q(M)$ is obvious. Therefore $Q(M)$ is a subvariety of $\mathbf{P}^5$. Since $\tilde{Q}(M)$ is evidently a 4-fold, the dimension of $Q(M)$ is at most 4. If it were less than 4, then every point of $Q(M)$ would have to have a preimage of dimension ≥ 1 in $\tilde{Q}(M)$. In particular, if $y \in Q(M)$, there would be two points x, z of $\Lambda(Q)$ such that (x, y) and (z, y) both lie in $\tilde{Q}(M)$, from which it would follow that x and z are joined by a line of the second kind in $\Lambda(M)$ and that $\mathcal{L}(M) = \tilde{Q}(M)$. Since the set of lines of the second kind is, by Lemma ???, an algebraic subset of dimension ≤ 1 of the Fano surface $F(M)$, the locus $\mathcal{L}(M)$ can have dimension at most 2, which is a contradiction. Therefore, $Q(M)$ is a hypersurface in $\mathbf{P}^4$ and the restriction of π_2 to the complement of $\mathcal{L}(M)$ is a bijection onto its image. That, however, is not quite the same as showing that it is an isomorphism. We also have to prove that the differential $d\pi_2$ is injective on the complement of $\mathcal{L}(M)$. The following argument accomplishes that.

Suppose (x, y) is a point of $\tilde{Q}(M)$ where the differential of π_2 vanishes. We can identify the tangent space to $\tilde{Q}(M)$ at (x, y) with the right null space of the matrix $(N_y \, M_x)$ modulo $(x \oplus y)$, where for indeterminates

$$X = (X_0, X_1, X_2, X_3, X_4)$$
$$Y = (Y_0, Y_1, Y_2, Y_3, Y_4, Y_5)$$

we define the matrix 5×6 matrix N_Y by

$$N_Y X = M_X Y.$$

The kernel of $d\pi_2$ can therefore be identified with

$$\mathbf{C}^5 \times \{0\} \cap ker(N_y \, M_x)$$

modulo $x \oplus \{0\}$. The latter can be identified, modulo x, with the set of all z in $\mathbf{C}^5$ such that $N_y z = 0$, or what is the same, $M_z y = 0$. Therefore, $d\pi_2$ fails to be injective at (x, y) if and only if the kernel of N_y has has dimension > 1 or, what is the same, there is a $z \neq x$ in $\mathbf{P}^4$ such that $M_x y = M_z y = 0$. But then the line joining L is a line of the second kind in the cubic $\Lambda(M)$ and $y = \varpi(L)$. This proves that π_2 maps $\tilde{Q}(M) - \mathcal{L}(M)$ isomorphically onto its image in $Q(M)$.

It remains to determine the degree of $Q(M)$. For this we use Schubert calculus. We know that the locus Ξ of singular alternating forms on $\mathbf{C}^6$ has, as a resolution of singularities, the projectivized vector bundle $\tilde{\Xi} = \mathbf{P}(\bigwedge^2(\mathbf{C}^6/E)^*)$, where E is the tautological 2-plane bundle over the Grassmannian of 2-planes in $\mathbf{C}^6$. The points of $\tilde{\Xi}$ may be identified with the set of pairs $([A], L)$ where A is a nonzero alternating form on $\mathbf{C}^6$ and L is a 2-plane in the kernel of A. The desingularization of Ξ is the mapping $\tilde{\Xi} \to \Xi$ given by $([A], L) \mapsto [A]$. It is an isomorphism over the open subset of Ξ consisting of all $[A]$ for which A has rank 4. The mapping $([A], L) \mapsto L$ projects $\tilde{\Xi}$ onto the Grassmannian $Gr = Gr(\mathbf{C}^2, \mathbf{C}^6)$.

We observe that M corresponds to a generic linear subvariety L_4 of $\mathbf{P}^{14}$ and is therefore the intersection of 10 generic hyperplanes. It is therefore Poincaé dual to a cohomology class on Ξ whose preimage in $\tilde{\Xi}$ is t^{10}, where t is the Chern class of the pullback of $\mathcal{O}_{\mathbf{P}^{14}}(1)$ to $\tilde{\Xi}$. On the other hand, the cohomology ring of $\tilde{\Xi}$ is $H^*(Gr)[t] = H^*(Gr)[T]/(p(T))$, where T is an indeterminate, where

$$p(T) = T^6 + b_1 T^5 + b_2 T^4 + b_3 T^3 + b_4 T^2 + b_5 T + b_6$$

and where the b_i are the Chern classes of the vector bundle $\bigwedge^2(\mathbf{C}^6/E)^*$ on Gr. The class t^{10} can therefore be written in the form

$$t^{10} = a_{10} + a_9 t + a_8 t^2 + a_7 t^3 + a_6 t^4 + a_5 t^5$$

where each a_i is a cohomology class of dimension i in the cohomology ring $H^*(Gr)$ of Gr. It corresponds to the locus in $\tilde{\Xi}$ of all pairs $([M_x], ker(M_x))$.

If we want to know how many of the lines $\mathbf{P}(ker(M_x))$ meet a generic line in $\mathbf{P}^5$, we must multiply the class t^{10} by an appropriate cohomology class in the Grassmannian, namely the class η corresponding to the Schubert cycle of all lines in $\mathbf{P}^5$ which meet a given line. That Schubert cycle has dimension 5 and the Grassmannian has dimension 8, so η is a 3-dimensional class. The product $t \cdot \eta$ is therefore of the form

$$t \cdot \eta = a_5 \eta t^5,$$

since the product $a_i \eta$ vanishes when $i > 5$. We are therefore reduced to computing the product $a_5 \eta$ in $H^*(Gr)$. To do that, we first have to identify η and a_5 in a form suitable for computation.

Following the notation of [Fu1], pp.271ff, the class η is given by

$$\eta = \{3, 0\}.$$

As for computing a_5, we proceed as follows. First, the total Chern class of E is given by

$$c(E) = 1 - \{1, 0\} + \{1, 1\}.$$

The total Chern class of $(\mathbf{C}^6/E)^*$ is therefore given by

$$c(\mathbf{C}^6/E) = 1 - \{1,0\} + \{2,0\} - \{3,0\} + \{4,0\}.$$

In order to obtain the total Chern class of $\bigwedge^2(\mathbf{C}^6/E)^*$ we need to introduce the formal factorization

$$T^4 - \{1,0\}T^3 + \{2,0\}T^2 - \{3,0\}T + \{4,0\} = \prod_{i=1}^{4}(T - \alpha_i),$$

in terms of which the total Chern class of $\bigwedge^2(\mathbf{C}^6/E)^*$ is given by

$$\prod_{1 \le i < j \le 4} (1 - \alpha_i - \alpha_j) = 1 + b_1 + b_2 + b_3 + b_4 + b_5 + b_6.$$

Using the computer algebra package REDUCE 3.4 on a personal computer, we find that

$$b_1 = -3\{1,0\} = -3\{1,0\}$$
$$b_2 = 3\{1,0\}^2 + 2\{2,0\} = 3\{1,1\} + 5\{2,0\}$$
$$b_3 = -\{1,0\}^3 - 4\{1,0\}\{2,0\} = -6\{2,1\} - 5\{3,0\}$$
$$b_4 = 2\{1,0\}^2\{2,0\} + \{2,0\}^2 + \{1,0\}\{3,0\} - 4\{4,0\} = 3\{2,2\} + 6\{3,1\}$$
$$b_5 = -\{1,0\}\{2,0\}^2 - \{1,0\}^2\{3,0\} + 4\{1,0\}\{4,0\} = -3\{3,2\}$$
$$b_6 = \{1,0\}\{2,0\}\{3,0\} - \{1,0\}^2\{4,0\} - \{3,0\}^2 = 0.$$

We then use the relation

$$t^6 = -b_1 t^5 - b_2 t^4 - b_3 t^3 - b_4 t^2 - b_5 t - b_6$$

to compute t^{10} and we find, again using REDUCE 3.4 on a personal computer,

$$t^{10} = (4\{4,1\} + 10\{3,2\})t^5 - (10\{3,3\} + 35\{4,2\})t^4 + 70\{4,3\}t^3 + 135\{4,4\}t^2.$$

In particular,

$$a_5 = 4\{4,1\} + 10\{3,2\}.$$

We therefore have

$$a_5\eta = (4\{4,1\} + 10\{3,2\}) \cdot \{3,0\} = 4\{4,4\}.$$

Therefore the locus $Q(M)$ is a quartic hypersurface and we are done.

Corollary (50.2): *In the special case where M is given by (47.1), the quartic hypersurface $Q(M)$ is defined by the following quartic*

$$\begin{aligned}(50.3) \qquad 0 = {}& y_0^4 + y_0(y_1^2 y_2 + y_2^2 y_3 + y_3^2 y_4 + y_4^2 y_5 + y_5^2 y_1) \\ & + y_1^2 y_3 y_5 + y_2^2 y_4 y_1 + y_3^2 y_5 y_2 + y_4^2 y_1 y_3 + y_5^2 y_2 y_4,\end{aligned}$$

This is the unique quartic invariant for an irreducible representation of dimension 6 of $SL_2(\mathbf{F}_{11})$.

Proof: The equivariance with respect to $SL_2(\mathbf{F}_{11})$ of the construction of (47.1) implies that the resulting quartic must be an invariant of $SL_2(\mathbf{F}_{11})$. A character computation shows that there is only one quartic invariant. By averaging the quartic y_0^4 over the group $SL_2(\mathbf{F}_{11})$ using the explicit matrices coming from the Weil representation, one obtains the above quartic. Both the character computation and the averaging are feasible to do by hand, but the author did it using REDUCE 3.4 on a personal computer.

§51 Generalization of Klein's A-curve of level 11

Klein showed that the modular curve $X(p)$ could be $PSL_2(\mathbf{F}_p)$ equivariantly mappted to the projective space $\mathbf{P}(V^+)$ associated to the even part V^+ of the Weil representation of $SL_2\mathbf{F}_p$). The resulting curve, which is called the *A*-**curve of level p** and denote $A(p)$, has the same genus as $X(p)$, namely

$$1 + \frac{(p^2 - 1)(p - 6)}{24},$$

and has degree

$$\frac{(p^2 - 1)(p - 1)}{48}.$$

Thus, for $p = 11$, the degree is 25 and the genus 26. In §§20,24 of [A-R], Ramanan and the author considered this embedding of $X(p)$ from our own point of view, but did not study the geometry of the image. In particular, we did not obtain the equations of $A(p)$.

Klein himself showed how to express the *A*-curve in terms of the *z*-curve in $\mathbf{P}(V^-)$. Denoting by $A_0, A_1, \ldots, A_{\frac{p-1}{2}}$ the coordinates of the $A(p)$ and by $z_1, \ldots, z_{\frac{p-1}{2}}$ those of the *z*-curve, Klein showed that

$$\frac{A_\alpha}{A_0} = \frac{z_{2\alpha}}{z_\alpha},$$

where for $1 \le \nu \le (p-1)/2$ we write $z_{p-\nu}$ to denote $-p_\nu$. We should however point out that in the treatment in [A-R], we failed to derive this relation. It would be desirable to do so from the point of view of [A-R].

Be that as it may, these relations do not provide explicit generators for the ideal of relations among the A_α alone. And in general, no such generators are known for special values of p, notably $p \ne 5, 7$. This is in stark contrast to the situation for the *z*-curve, whose equations were given explicitly by Klein.

In this section, we give explicit equations defining Klein's A-curve $A(11)$ of level 11. This result was stated, without proof, for the first time in the author's article [A13]. In this section, we will give a proof. The result is as follows.

Theorem (51.1): *The A-curve $A(11)$ of level 11 is the*   *singular locus of the quartic hypersurface in $\mathbf{P}^4$ given by (50.3).*

We will give part of the proof now and the rest of it later on in this section, after some general considerations of the singular loci of the quartics $Q(M)$ for sufficiently generic M in $\mathcal{P}$. Since the curve has degree 25, if it does not lie in the quartic, it meets the quartic in 100 points, with multiplicity. The orbits of $PSL_2(\mathbf{F}_{11})$ on $X(11)$ have orders 60, 220, 330, and 660, but it is not possible to express 100 as an integral linear combination of these numbers with nonnegative coefficients. The curve $A(11)$ therefore lies on this quartic.

The curve $A(11)$ is embedded via the line bundle α, which is uniquely determined by its degree 25 and by its $SL_2(\mathbf{F}_{11})$-invariance. Since $X(11)$ has genus 26, it follows that α is a semicanonical line bundle. Cubics in $\mathbf{P}^5$ therefore cut out a sesquicanonical linear system on $A(11)$ corresponding to the line bundle α^3. By the Riemann-Roch theorem, the space $\Gamma(\alpha^3)$ of sections of α^3 has dimension $3 \cdot 25 + 1 - 26 = 50$, while the dimension of the space $Sym^3(\Gamma(\alpha))$ of cubic forms in 6 variables is 56. Therefore, if we believe that the natural mapping from $Sym^3(\Gamma(\alpha)) \to \Gamma(\alpha^3)$ is surjective, the kernel will be a 6 dimensional space of cubics and it is then easy to believe that they are the 6 partials of our quartic. We don't know how to proceed with that argument, but we can nevertheless prove that these 6 partials vanish on $A(11)$ by proving that the 6 dimensional representation of $SL_2(\mathbf{F}_{11})$ on V^+ does not occur in the representation of $SL_2(\mathbf{F}_{11})$ on $\Gamma(\alpha^3)$. That can be done by a straightforward use of the Woods Hole fixed point formula, which we now describe briefly. Since $H^1(\alpha^3)$ vanishes, the character χ of $SL_2(\mathbf{F}_{11})$ on $\Gamma(\alpha^3)$ is given by

$$\chi(g) = \begin{cases} \pm 50 & \text{if } g = \pm 1 \\ \sum_{g \cdot x = x} \frac{d\alpha_x^3(g)}{1 - dg_x} & \text{if not,} \end{cases}$$

where in the second case the summation runs over all of the fixed points x of g on the modular curve $X(11)$, where $\alpha_x(g)$ is the eigenvalue of g on the fibre of α at x, and where dg_x is the eigenvalue of g on the fibre of the tangent bundle of $X(11)$ at x. The number we wish to compute is

$$\frac{1}{\circ(SL_2(\mathbf{F}_{11}))} \sum_g \overline{\rho_6(g)} \chi(g),$$

where g runs over all of the elements of the group $SL_2(\mathbf{F}_{11})$ and where ρ_6 is the 6-dimensional representation we wish to prove does not occur. Since we have $\chi(-g) = -\chi(g)$ and $\rho_6(-g) = -\rho_6(g)$, the contribution of g to the sum is the same as that of $-g$. Therefore, the sum may be reduced to a sum running over $PSL_2\mathbf{F}_{11}$), namely

$$\frac{1}{\circ(PSL_2(\mathbf{F}_{11}))} \sum_g \overline{\rho_6(\bar{g})} \chi(\bar{g}),$$

where g now runs over $PSL_2(\mathbf{F}_{11})$ and $\bar{g}$ denotes a representative of g in $SL_2(\mathbf{F}_{11})$. Since an element of $PSL_2(\mathbf{F}_{11})$ cannot have a fixed point unless its order is 1, 2, 3 or 11, the character $\chi(g)$ vanishes unless g is such an element. On the other hand, the character $\rho_6(g)$ vanishes if g has order 2 or 3. Therefore, the only contributions to the above summation come from the identity element and the elements of order 11. Finally since each term of the summation depends only the conjugacy class of

g, the multiplicity of ρ_6 in χ is

$$\frac{1}{660}\left[6\cdot 50 + 60\left(\frac{1-\sqrt{-11}}{2}\right)Tr\left(\frac{\zeta_{11}^7}{1-\zeta_{11}}\right) + 60\left(\frac{1+\sqrt{-11}}{2}\right)Tr\left(\frac{\zeta_{11}^4}{1-\zeta_{11}^2}\right)\right],$$

where Tr denotes the trace from the cyclotomic field $\mathbf{Q}(\zeta_{11})$ to the quadratic field $\mathbf{Q}(\sqrt{-11})$. We evaluate this and find that it vanishes. (A similar computation was done in §24 of [A-R]).

Alternatively, we can use the explicit theta series given by Klein in [K2] to embed $X(11)$ as the A-curve $A(11)$ and to verify directly that the first partials of the quartic vanish when these series are substituted for the variables. The series used by Klein to embed $X(p)$ as the A-curve $A(p)$ are given (cf. [K3], p.188, eq.(5)) by

$$\rho A_\alpha = \begin{cases} \sum_{\lambda=-\infty}^{\infty}(-1)^\lambda \cdot q^{\frac{(6\lambda+1)^2 p}{12}} & \text{if } \alpha = 0 \\ (-1)^\alpha \cdot \sum_{\lambda=-\infty}^{\infty}(-1)^\lambda \cdot \left\{q^{\frac{((6\lambda+1)p+6\alpha)^2}{12p}} + q^{\frac{((6\lambda+1)p-6\alpha)^2}{12p}}\right\}, & \text{if not} \end{cases}$$

in general. One can compute these expansions for the special case $p = 11$ or find that on p.161 of [K2] Klein has already done so (cf. eq. (30)). One then verifies that if we put

$$x_0 = A_0$$
$$x_1 = A_2$$
$$x_2 = A_5$$
$$x_3 = A_4$$
$$x_4 = A_1$$
$$x_5 = A_3$$

then all 6 first partials of the quartic vanish. Using the computer algebra package REDUCE 3.4 on a personal computer, the author has verified this up to order 10,000 in q, with the series computed by taking the summations from $\lambda = -100$ to $\lambda = 100$.

This completes the first part of the proof of Theorem 51.1, namely, we have shown that the A-curve of level 11 lies in the singular locus of the quartic (50.3).

Notation (51.2):   *Let M be a sufficiently generic element of $\mathcal{P}$. We will denote by $\Sigma(M)$ the image of $\mathcal{L}(M)$ in $Q(M)$ under π_2.*

Lemma (51.3):   *Let M be a sufficiently generic element of $\mathcal{P}$. $\Sigma(M)$ is isomorphic to the locus of jumping lines of $E(M)$ in the Fano surface $F(M)$ of $\Lambda(M)$.*

Proof: We have everywhere defined morphisms in both directions and they are easily seen to be inverses of each other. If L is a jumping line then the corresponding point of $\Sigma(M)$ is $\varpi(L)$. Conversely, if y is a point of $\Sigma(M)$ then the preimage of y under π_2 is, by definition of $\Sigma(M)$, a line of the third kind $L^{\#}$ whose projection into $\Lambda(M)$ under π_1 is a jumping line L.

Corollary (51.4):   *Let M be a sufficiently generic element of $\mathcal{P}$. The locus $\Sigma(M)$ has dimension at most 1 and the locus $\mathcal{L}(M)$ has dimension at most 2.*

Proof: The first assertion follows from the preceding lemma and from Corollary 49.9. The second follows from the fact that the fibres of $\mathcal{L}(M)$ over $\Sigma(M)$, being projective lines, have dimension 1.

Lemma (51.5):   Let M be a sufficiently generic element of $\mathcal{P}$. Then $\Sigma(M)$ is the singular locus of $Q(M)$.

Proof: I am indebted to Yum-Tong Siu for the following argument. Let y be a point of $\Sigma(M)$ and suppose that y is not a singular point of $Q(M)$. Then we can find a nowhere vanishing holomorphic 4-form ω defined in a (classical) neighborhood U of y. Denote by V the preimage of U in $\tilde{Q}$ under π_2 and let $\eta = \pi_2^* \omega$ be the pull back of ω. Since the fibres of π_2 over points of $\Sigma(M)$ have positive dimension, η is a holomorphic 4-form on V which vanishes on $V \cap \mathcal{L}(M)$ and nowhere else. But that is a contradiction, since the locus where a holomorphic 4-form vanishes on a complex 4-manifold must be a divisor, whereas $\mathcal{L}(M)$, by the preceding corollary, has dimension at most 2. This proves that $\Sigma(M)$ coincides with the singular locus of $Q(M)$.

Lemma (51.6):   The singular locus $\Sigma(M)$ of $Q(M)$ contains a curve.

Proof: Suppose not. Then the singular locus consists entirely of isolated points, each of which must arise by collapsing a line of the third kind in $\tilde{Q}(M)$ to a point. Using a Mayer-Vietoris argument on $\tilde{Q}(M)$, where a tubular neighborhood U of a line of the third kind $L^\#$ retracts to $L^\#$, and on $Q(M)$, where the image of such a neighborhood is contractible, we see that the third singular homology group with real coefficients of $\tilde{Q}(M)$ is mapped isomorphically onto that of $Q(M)$. However, it is easy to compute the third homology of $\tilde{Q}(M)$ using the Leray-Hirsch theorem and the known homology of the cubic threefold $\Lambda(M)$. One concludes that the homology is not zero. In fact, it has dimension 10. On the other hand, it follows from the Lefshetz theorem on hyperplane sections (cf. [Mi], Corollary 7.3, p.41)[28] that the third homology of $Q(M)$ is the same as that of $\mathbf{P}^5$ and therefore vanishes. That is a contradiction and the lemma is proved.

Lemma (51.7):   Let M be a sufficiently generic element of $\mathcal{P}$. Then the quartic hypersurface $Q(M)$ has no isolated singular points.

Proof: I am indebted to Torsten Ekedahl and Gabor Megyesi for their help with this argument. Suppose y is an singularity of $Q(M)$. Then by Lemma 51.5, y lies in $\Sigma(M)$ and the preimage of y in $\tilde{Q}(M)$ under π_2 is a line of the third kind $L^\#$. Let $L = \pi_1(L^\#)$ be the line of the second kind in $\Lambda(M)$ associated to $L^\#$. According to Prop. 6.19 of [C-G], the normal bundle of L in Λ is isomorphic either to $\mathcal{O} \oplus \mathcal{O}$ or to $\mathcal{O}(-1) \oplus \mathcal{O}(1)$. Therefore, the normal bundle of $L^\#$ in $\tilde{Q}(M)$ is either $\mathcal{O}(-2) \oplus \mathcal{O} \oplus \mathcal{O}$ or to $\mathcal{O}(-2) \oplus \mathcal{O}(-1) \oplus \mathcal{O}(1)$. In either case, denoting by N the normal bundle of $L^\#$ in $\tilde{Q}(E)$, we have $h^0(N) = 2$ and $h^1(N) = 1$. Therefore, the deformation space of $L^\#$ in $\tilde{Q}(M)$ is a curve and we can deform $L^\#$ in $\tilde{Q}(M)$ as one of the fibres of a ruled surface S lying in $\tilde{Q}(M)$. All of the curves obtained from $L^\#$ by deformation will also be collapsed. Since the preimage of y is $L^\#$ and not a surface, the image of S in $Q(M)$ will be a curve. By Lemma 51.5, that curve must lie in the singular locus of $Q(M)$, which proves that y is not an isolated singularity.

---

[28] Thanks to Torsten Ekedahl for providing this reference.

Lemma (51.8):   *Let M be as in (46.1). For $0 \leq i \leq 5$, denote by e_i the point of $\mathbf{P}^5$ whose only nonzero coordinate is the i-th coordinate (e.g. $e_1 = [0,1,0,0,0,0]$). Then the 5 points e_1, e_2, e_3, e_4, e_5 are simple points of $\Sigma(M)$. In particular, there is exactly one irreducible component of $\Sigma(M)$ through each of these points.*

Proof: The second assertion follows from the first and from the preceding lemma. Since $Q(M)$ is invariant under the cyclic permutation $(y_1 y_2 y_3 y_4 y_5)$, which permutes these 5 points transitively, it will be enough to prove that $[0,1,0,0,0,0]$ is a simple point of $Q(M)$. Since $\Sigma(M)$ is defined by the 6 partials of (50.3), the singular locus of $\Sigma(M)$ is the locus of points y of $\mathbf{P}^5$ where the matrix of second partials of (50.3) has rank < 4. But if we compute that matrix at the point $[0,1,0,0,0,0]$, we find that it is

$$\begin{pmatrix} 0 & 0 & 1 & 0 & 0 & 0 \\ 0 & 0 & 0 & 0 & 0 & 0 \\ 1 & 0 & 0 & 0 & 0 & 0 \\ 0 & 0 & 0 & 0 & 0 & 1 \\ 0 & 0 & 0 & 0 & 0 & 0 \\ 0 & 0 & 0 & 1 & 0 & 0 \end{pmatrix}$$

which has rank exactly 4. This proves the lemma.

Lemma (51.9):   *Let M be the matrix (47.1). For $0 \leq i \leq 5$, let e_i be as in Lemma 51.7. Then the hyperplane $x_0 = 0$ meets $\Sigma(M)$ in precisely 5 points, namely e_1, e_2, e_3, e_4, e_5. Furthermore, the hyperplane $x_0 = 0$ meets the curve $\Sigma(M)$ with multiplicity 5 at each of these points.*

That the hyperplane $x_0 = 0$ meets $A(11)$ with multiplicity 5 at each of the points e_1, e_2, e_3, e_4, e_5 was shown by Klein ([K2], p.164, eq.(32)), this being the heart of his proof that the curve $A(11)$ has degree 25. One can probably also give a valuation theoretic proof as was done in §23 of [A-R] for Klein's z-curve, but we will not attempt to do so. We will therefore be done if we can prove that Σ is smooth at these 5 points and that the hyperplane $y_0 = 0$ does not meet Σ at any other points. In view of the invariance of the locus Σ and of the hyperplane $y_0 = 0$ under cyclic permutation of the coordinates y_1, y_2, y_3, y_4, y_5, it will suffice to show that e_1 is the only point of Σ where $y_0 = 0$ and $y_1 \neq 0$. That means we are in effect working in the affine open subset $y_1 \neq$ and can accordingly study the equations of Σ with the additional equations $y_0 = 0$ and $y_1 = 1$.

The first partials of the invariant quartic, which we will denote q_0, q_1, $\ldots$, q_5 respectively, are given by

$$q_0 = 4y_0^3 + y_1^2 y_2 + y_2^2 y_3 + y_3^2 y_4 + y_4^2 y_5 + y_5^2 y_1$$

$$q_1 = y_0(2y_1 y_2 + y_5^2) + 2y_1 y_3 y_5 + y_2^2 y_4 + y_4^2 y_3$$

$$q_2 = y_0(2y_2 y_3 + y_1^2) + 2y_2 y_4 y_1 + y_3^2 y_5 + y_5^2 y_4$$

$$q_3 = y_0(2y_3 y_4 + y_2^2) + 2y_3 y_5 y_2 + y_4^2 y_1 + y_1^2 y_5$$

$$q_4 = y_0(2y_4 y_5 + y_3^2) + 2y_4 y_1 y_3 + y_5^2 y_2 + y_2^2 y_1$$

$$q_5 = y_0(2y_5 y_1 + y_4^2) + 2y_5 y_2 y_4 + y_1^2 y_3 + y_3^2 y_2$$

and when we put $y_0 = 0$, $y_1 = 1$ and denote the resulting value of q_i by r_i, we have

$$r_0 = y_2 + y_2^2 y_3 + y_3^2 y_4 + y_4^2 y_5 + y_5^2$$
$$r_1 = 2y_3 y_5 + y_2^2 y_4 + y_4^2 y_3$$
$$r_2 = 2y_2 y_4 + y_3^2 y_5 + y_5^2 y_4$$
$$r_3 = 2y_3 y_5 y_2 + y_4^2 + y_5$$
$$r_4 = 2y_4 y_3 + y_5^2 y_2 + y_2^2$$
$$r_5 = 2y_5 y_2 y_4 + y_3 + y_3^2 y_2$$

Our job is to show that these equations imply that $y_2 = y_3 = y_4 = y_5 = 0$.

First we note that if y_2 or y_4 is 0, i.e. if $y_2 y_4 = 0$, then r_5 becomes $y_3(y_2 y_3 + 1)$, hence $0 = y_4 r_5 = y_3 y_4$. Therefore $0 = r_1 = 2y_3 y_5$, so $y_3 y_5 = 0$. Therefore $0 = r_2 = y_4^2 y_5$, so $y_4 y_5 = 0$. We then have $0 = r_3 = y_4^2 + y_5$, which together with $y_4 y_5 = 0$ implies $y_4 = y_5 = 0$. This implies $0 = r_4 = y_2^2$, so $y_2 = 0$, and $0 = r_5 = y_3$. Therefore, $[y_0, y_1, y_2, y_3, y_4, y_5] = [0, 1, 0, 0, 0, 0]$, as promised.

So we will be done if we can show that $y_2 y_4 = 0$. Hence suppose $y_2 y_4 \neq 0$. We can then solve for y_5^2 in $0 = r_4$ and obtain

$$y_5^2 = -\frac{y_2^2 + 2y_3 y_4}{y_2}.$$

Let s_2 be the numerator of r_2 after this substition and obtain

$$s_2 = y_2^3 + y_3 y_4^2 + 2y_3 y_5.$$

Subtracting s_2 from r_1 to eliminate the common term $y_2^2 y_4$, we obtain $y_3(-y_2 y_3 y_5 + 3y_4^2 + 2y_5)$. We cannot have $y_3 = 0$ since then $r_1 = 0$ becomes $y_2^2 y_4 = 0$, contradicting our assumption that $y_2 y_4 \neq 0$. Therefore $y_3 \neq 0$ and the expression $-y_2 y_3 y_5 + 3y_4^2 + 2y_5$, which we denote s_{12} must vanish. Eliminating the term $y_2 y_3 y_5$ between s_{12} and r_3, we have

$$0 = 2s_{12} + r_3 = 7y_4^2 + 5y_5,$$

which lets us solve for y_5, namely $y_5 = -\frac{7}{5} y_4^2$. Since $y_4 \neq 0$ and the numerator of r_3 is $-2y_4^2(7y_2 y_3 + 1)$, we conclude from $r_3 = 0$ that $y_2 y_3 = -\frac{1}{7}$. The numerator of r_1 then becomes $y_4(5y_2^2 - 9y_3 y_4)$, which implies $5y_2^2 - 9y_3 y_4$. Multiplying this expression by y_2 and using $y_2 y_3 = -\frac{1}{7}$, we have $5y_2^3 - \frac{9}{7} y_4 = 0$, which lets us solve for y_4:

$$y_4 = \frac{35}{9} y_2^3.$$

After this, the numerator of q_2 becomes $10y_2^2$, hence $y_2 = 0$, which contradicts our assumption $y_2 y_4 \neq 0$. This completes the proof of the lemma.

Corollary (51.10): *Let M be given by (47.1). Then the curve* $\Sigma(M)$ *is an irreducible curve.*

Proof: Any component of $\Sigma(M)$ must meet the hyperplane $y_0 = 0$. By Lemma 51.8, the only intersections of y_0 with $\Sigma(M)$ are at the 5 points e_i, with $1 \leq i \leq 5$. We know from Lemma 51.7 that there is exactly one component of $\Sigma(M)$ passing through each of these points. But Klein's A-curve $A(11)$ passes through all five of the points e_i with $1 \leq i \leq 5$. Therefore, it is the only component of $\Sigma(M)$, which proves that $\Sigma(M)$ is irreducible.

Remark (51.11): With the preceding corollary, the proof of Theorem 51.1 is now complete.

Corollary (51.12): *Let M be a sufficiently generic element of $\mathcal{P}$. Then the singular locus $\Sigma(M)$ of the quartic hypersurface $Q(M)$ is an irreducible curve.*

Proof: We know from Lemmas 51.4, 51.5 and 51.6 that the singular locus $\Sigma(M)$ of $Q(M)$ is a curve. Since irreducibility of $\Sigma(M)$ is an open condition on M, it suffices to show that in the special case of (47.1), the curve $\Sigma(M)$ is irreducible. But that has already been done in the preceding corollary.

Lemma (51.13): *Let M be given by (47.1). Then the curve $\Sigma(M)$ is a smooth curve of degree 25 and genus 26.*

Proof: We know from Corollary 51.1 that the A-curve $A(11)$ of level 11 is the singular locus of $Q(M)$. Since the mapping of $X(11)$ onto $A(11)$ is equivariant for $PSL_2(\mathbf{F}_{11})$, which acts irreducibly and nontrivially on $\mathbf{P}^5$, the image of $X(11)$ has arithmetic genus 26. That it has degree 25 was shown by Klein [K3]. It remains to show that $\Sigma(M)$ is nonsingular. Since the A-curve is invariant under the action of the group $PSL_2(\mathbf{F}_{11})$, its singular locus is a union of orbits for that group. Therefore, if $A(11)$ is singular, it must be singular along an entire orbit for $PSL_2(\mathbf{F}_{11})$. The only subgroups of $PSL_2(\mathbf{F}_{11})$ of index < 60 are conjugate either to the group B of upper triangular matrices, which has index 12, or to the alternating group A_5 on 5 letters, which has index 11. Since the restriction of the even part of the Weil representation to A_5 is irreducible, there is no orbit of order 11. Since the only fixed point of the group B is the point $[1, 0, 0, 0, 0, 0]$, which does not lie on the quartic hypersurface and therefore not on $A(11)$, there is no orbit of order 12 on $A(11)$. Therefore, if $A(11)$ has a singular point, it must have at least 60 of them. Denote the precise number of singular points by S. We can project the $A(11)$ into $\mathbf{P}^3$ from a generic line in $\mathbf{P}^5$ to obtain a curve $A^\flat(11)$ of degree 25 and genus 26 in $\mathbf{P}^3$ with at least S singularities. If we then project from a generic point of $A'(11)$ into a plane, we obtain a plane curve $A''(11)$ of degree $d = 24$ and genus $g = 26$ with at least S singularities. By the Plücker formula, the number of singularities of $A''(11)$, counted with their multiplicity, is $(d-1)(d-2)/2 - g = 253 - 26 = 217 < 220$, so $S < 220$. Since the next size larger than 60 that a $PSL_2(\mathbf{F}_{11})$ orbit can have[29] on $A(11)$ is 220, it follows that if $A(11)$ is singular, it must have exactly 60 singularities coinciding with the unique 60 point orbit of $PSL_2(\mathbf{F}_{11})$ in $\mathbf{P}^5$. We will now show that this cannot happen.

   Since $S \geq 60$, choose 38 of the singular points $p_1, \ldots, p_{38}$. If K is a cubic hypersurface passing through these 38 points then these 38 intersections, being singular on $A(11)$, count for $2 \times 38 = 76$ intersections in all. But by Bezout's

---

[29]   The point here is that after 60, the next larger index that a subgroup of $PSL_2(\mathbf{F}_{11})$ can have is 220.

theorem, we can only have $deg(A(11)) \times deg(K) = 3 \times 25 = 75$ intersections unless K contains $A(11)$. It follows that every cubic through these 38 points must contain the A-curve. Hence the A-curve lies in at least $55 - 38 = 17$ linearly independent cubic hypersurfaces, of which 6 are accounted for by the partials of the invariant quartic. Denote by W the space of all cubics vanishing on the A-curve. It is straightforward to show by a character computation that the representation of $SL_2(\mathbf{F}_{11})$ on cubic forms on $\mathbf{C}^6$ decomposes into the sum of two components of degree 12, two of degree 10 and one each of the two irreducible representations of degree 6. One of these two irreducible representations of degree 6 arises from the first partials of the invariant quartic and the eigenvalues of the element $\begin{pmatrix} 1 & 1 \\ 0 & 1 \end{pmatrix}$ in that 6 dimensional representation are ζ_{11}^2, ζ_{11}^6, ζ_{11}^7, ζ_{11}^8, ζ^{10}. In particular, the eigenvalue ζ_{11} does not occur. On the other hand, the eigenvalue ζ_{11} does occur in the other irreducible representation of degree 6 and in all irreducible representations of degrees 10 or 12. Therefore, we will be done if we can show that there is no cubic with eigenvalue ζ which vanishes on the 60 point orbit. If we enumerate the 56 monomials of degree 3 and examine their eigenvalues, we find that the only monomials that have eigenvalue ζ_{11} are $x_0^2 x_1$, $x_0 x_3 x_5$, $x_2 x_4 x_5$, x_2^3, $x_3^2 x_4$, so the most general cubic with eigenvalue ζ_{11} is

$$a_0 x_0^2 x_1 + a_1 x_0 x_3 x_5 + a_2 x_2 x_4 x_5 + a_3 x_2^3 + a_4 x_3^2 x_4.$$

Since the 60 point orbit contains $[0, 0, 1, 0, 0, 0]$, we have $a_4 = 0$. We can also evaluate this cubic at other points of the orbit to impose more conditions. Using REDUCE 3.4 on a personal computer, the author verified that these conditions are linearly independent. Therefore, there is no cubic of eigenvalue ζ_{11} which vanishes on the 60 point orbit, hence the A-curve is nonsingular.

Corollary (51.14):   *Let M be a sufficiently generic element of $\mathcal{P}$. Then the curve $\Sigma(M)$ is a smooth curve.*

Proof: Smoothness of $\Sigma(M)$ is an open condition on M. Therefore we just need to know that the set U of such M is nonempty. But by the preceding lemma, U contains (47.1), so we are done.

Lemma (51.15):   *Let M be a sufficiently generic element of $\mathcal{P}$. Then the curve $\Sigma(M)$ is a smooth curve of degree 25 and genus 26.*

Proof: That the degree is 25 follows from conservation of number ([Fu1], Cor.10.2.2, p.182) and the fact that the degree of Klein's A-curve is 25. The rest of the proof is the same, mutatis mutandis, as the proof of Proposition 46.4.

We take this opportunity to endow the locus $\Sigma(M)$ with a name.

Definition (51.16):   *Let M be a sufficiently generic element of $\mathcal{P}$. Then we will call $\Sigma(M)$ the **A-curve of M**. We will also refer to it as **an A-curve of $\Lambda(M)$**.*

§52   The Jacobian variety of an A-curve of a generic cubic threefold

In this section, we prove the following result.

Theorem (52.1):   *Let M be a sufficiently generic element of $\mathcal{P}$. Then the Jacobian variety of the A-curve of M is the sum of an abelian variety $\mathcal{A}_{21}(M)$ of dimension 21 and an abelian variety $\mathcal{A}_5(M)$ of dimension 5. Furthermore, $\mathcal{A}_5(M)$ is isogenous to the intermediate Jacobian of the cubic threefold $\Lambda(M)$ and is actually isomorphic to it in the case where M is given by (47.1).*

Proof: According to Lemma 51.3, there is a canonical isomorphism of $\Sigma(M)$ onto the locus of jumping lines of the vector bundle $E(M)$ on $\Lambda(M)$. Denote by κ the composition of that isomorphism with the inclusion of the locus of jumping lines into the Fano surface $F(M)$ of $\Lambda(M)$. Then κ induces a morphism from the Jacobian variety of the A-curve $\Sigma(M)$ to the Albanese variety of the Fano surface or, what is the same, to the intermediate Jacobian variety of the cubic threefold in the middle dimension. Therefore, we will be done if we can show that these morphisms of abelian varieties are surjective. Since surjectivity is an open condition on M, it will be enough to show that this is so for the $PSL(2, 11)$-invariant alternating form corresponding to Klein's cubic. In that case, the mapping between the abelian varieties is $PSL(2, 11)$ equivariant. Therefore, the image must be a $PSL(2, 11)$-invariant abelian subvariety of $Alb(F)$, hence either 0 or all of $Alb(F)$. Since the mapping κ is injective and since F injects into $Alb(F)$ and $\Sigma(M)$ into $J(\Sigma(M))$, the following commutative diagram shows that the image of $J(\Sigma(M))$ cannot be 0:

$$
\begin{array}{ccc}
\Sigma(M) & \rightarrow & F \\
\downarrow & & \downarrow \\
J(\Sigma(M)) & \rightarrow & Alb(F)
\end{array}
$$

This proves the theorem.

§53   The fundamental intertwining operator revisited

In §51 we showed how to generalize Klein's A-curve of level 11 to the context of cubic threefolds or, more precisely, of skew-symmetric 6×6 matrices whose entries are linear forms in 5 variables. The corresponding generalization of Klein's z-curve of level 11 has been known for some time and was probably understood by Klein himself. However, since no one has seen fit to name the generalization accordingly, we will do so now.

Definition (53.1):   *Let Λ be a sufficiently generic cubic threefold. Then the nodal curve of the Hessian of Λ will be called the z-**curve** of Λ.*

The motivation for this terminology comes from Klein's theorem that the singular locus of the Hessian of Klein's cubic threefold is isomorphic to the modular curve of level 11. More generally, Klein embedded the modular curve of level p $PSL_2(\mathbf{F}_p)$

equivariantly into a projective space of dimension $(p-3)/2$. The image was a curve of degree $(p^2 - 1)(p - 3)/48$ which he called the z-curve. In the case $p = 11$, the degree is 20 and it arises from Klein's cubic threefold in the manner just described. However, it is known (cf. e.g. [E1] and references therein) that the singular locus of the Hessian of a generic cubic threefold is likewise a smooth curve of degree 20 and genus 26. Therefore, the above definitition is perfectly justified.

The reader will recall that our discovery of the equations of Klein's A-curve of level 11 and its generalization to cubic threefolds began with the small observation that, using the fundamental intertwining operator of [A-R], the matrix (47.0) of second partials of Klein's cubic. could be transformed $SL_2(\mathbf{F}_{11})$ equivariantly into the 6×6 skew-symmetric matrix (47.1). Furthermore, Klein's z-curve is the rank 3 locus of (47.0) while Klein's A-curve is defined in terms of (47.1). However, the fundamental intertwining operator, which has only been defined in the case of Klein's cubic and at first sight only appears to make sense in that case, has not itself been generalized so far. In this section, we will suggest one way in which this might be accomplished.

Begin with a sufficiently generic matrix M in $\mathcal{P}$. Then M determines a projective 4-plane Π_M in the projective space $\mathbf{P}(\bigwedge)$ of skew-symmetric 6×6 matrices, this being a projective space of dimension 14. Denote by $\mathcal{P}f$ the cubic hypersurface in $\mathbf{P}^{14}$ consisting of singular matrices. For any point m of $\mathbf{P}^{14}$, the polar quadric of m with respect to $\mathcal{P}f$ will be denoted $\mathcal{Q}_m$ and depends linearly on m. In this way, we get a family of ∞^{14} quadrics in $\mathbf{P}^{14}$ indexed by $\mathbf{P}^{14}$ itself. If we now intersect each of these quadrics with Π_M, we obtain a family of ∞^{14} quadrics in Π_M. Since the family of all quadrics in Π_M is itself a projective space $\mathcal{Q}$ of dimension 14, we therefore have a linear mapping of $\mathbf{P}(\bigwedge)$ onto $\mathcal{Q}$. This is a tempting candidate for a generalization of the fundamental intertwining operator, but it remains to be seen whether this candidate deserves to be elected. One important issue in the election will be whether one can use the proposed generalization to obtain descriptions of some of the A-curves or z-curves associated to cubic threefolds analogous to the description found for the modular curve $X(p)$, i.e. as the intersection of a Grassmannian and a 2-uply embedded projective space.

References

[A1] Allan ADLER, "On the automorphism group of a certain cubic threefold," American Journal of Mathematics **100** (1978), 1275-1280

[A2] Allan ADLER, "On $V^2W+W^2X+X^2Y+Y^2Z+Z^2V = 0$ and Related Topics," preprint, Institute for Advanced Study, 1975

[A3] Allan ADLER, "On the Weil Representation," preprint, reproduced here as App. I.

[A4] Allan ADLER, "Modular Correspondences on $X(11)$," Proc. Edinburgh Math. Soc.(2) **35** (1992) 427-435

[A5] Allan ADLER, "Cubic Invariants for $SL_2(\mathbf{F}_q)$," J. Algebra **145** (1992) 178-186

[A6] Allan ADLER, "Invariants of $SL_2(\mathbf{F}_q)$ Acting on $\mathbf{C}^n$ for $q = 2n \pm 1$," to appear in an MSRI volume on the Klein curve

[A7] Allan ADLER, "Some Integral Representations of $PSL_2(\mathbf{F}_p)$ and Their Applications," J. of Algebra **72** (1981) 115-145

[A8] Allan ADLER, "On the ring of invariants of $PSL_2(\mathbf{F}_{11})$ Acting on $\mathbf{C}^5$," Preprint, Tata Institute, Bombay 1981

[A9] Allan ADLER, "Invariants of the Simple Group of Order 660 Acting on $\mathbf{C}^5$," Comm.Alg. **20** (1992) 2837-2862

[A10] Allan ADLER, "On the Automorphism Groups of Certain Hypersurfaces," J.Alg. **72** (1981) 146-165

[A11] Allan ADLER, "Invariants and $X(11)$," preprint, Tata Institute 1981. Reproduced here as App. III.

[A12] Allan ADLER, "Determination of Modular Correspondences via Geometric Properties," Pac.J.Math. **155** (1992) 1-27

[A13] Allan ADLER, "The Mathieu group M_{11} and the modular curve $X(11)$" Proc. London Math. Soc. **74** (1997)

[A14] Allan ADLER, "Modular forms of weight 4/5 for $\Gamma(11)$," preprint, reproduced as App. II to this volume.

[A15] "New Abelian Varieties Associated to Cubic Threefolds," preprint, reproduced here as App. V

[A16] "On the Hessian of a Cubic Threefold," preprint, reproduced here as App.IV

[A17] Allan ADLER, "On the automorphism groups of certain hypersurfaces II," Comm. Alg. **27** (1994) 2319-2366

[A18] Allan ADLER, "Notes of the Off Broadway Seminar," IHES Preprint, 1995.

[Ba] H.F. BAKER, A locus with 25920 self transformations, Cambridge Tracts in Mathematics and Mathematical Physics, no. 39, 1946

[Bi 1] Christina BIRKENHAKE, "Heisenberg Gruppen ampler Geradenbündel auf abelschen Varietäten," Doctoral Dissertation, Zürich, Switzerland, November 11,1989

[Bi-La] Christina BIRKENHAKE and Herbert LANGE, Complex Abelian Varieties, Grundlehren der mathematischen Wissenschaften 302, Springer-Verlag 1992

[BLR] Siegfried BOSCH, Werner LUTKEBÖHMERT, Michel RAYNAUD, *Neron Models*, Ergebnisse der Mathematik und ihrer Grenzgebiete, 3 Folge, Band 21, Springer-Verlag 1980

[Br] Richard BRAUER, "On groups whose order contains a prime to the first power only," American Journal of Mathematics **64** (1942) 401-440.

[Bri] Francesco BRIOSCHI, "Sulla trasformazione dell'undecimo ordine delle funzioni ellittiche," Annali di Matematica pura ed applicata, serie II,**21** (1893) 309-315 [Opere Matematiche, vol.3, pp.43-50 Milan: Ulrico Hoepli, 1904]

[Bu] Heinrich BURCKHARDT, "Untersuchungen aus dem Gebiete der hyperelliptischen Modulfunktionen III," Mathematische Annalen **41** (1893) 313-43

[Cl1] William Kingdon CLIFFORD, "On the classifocation of loci," Phil. Trans. Roy. Soc. **169** (1879) 663 [Mathatical Papers (London, 1882) 305-331]

[Co1] Arthur COBLE, "Point Sets and Allied Cremona Groups III," Transactions of the American Mathematical Society **18** (1917) 331-372

[Co2] Arthur COBLE, *Algebraic Geometry and Theta Functions*, American Mathematical Society Colloquium Publications, Volume X, 1969 American Mathematical Society, Providence, Rhode Island 02904

[Con] J.H.CONWAY, R.T.CURTIS, S.P.NORTON, R.A.PARKER, R.A.WILSON, *Atlas of Finite Groups*, *Clarendon Press*, Oxford 1985

[C-G] Herb CLEMENS and Philip GRIFFITHS, "The intermediate jacobian of a cubic threefold," Annals of Mathematics **95** (1972) 281-356

[D] L. DORNHOFF, *Group Representation Theory*, Dekker, New York, 1971-2.

[De-Ra] Pierre DELIGNE and M.RAPAPORT, "Les schémas de modules de courbes elliptiques," pp.143-316 of Modular Functions of One Variable II, Proc. of Int'l Summer School, University of Antwerp, July 17-Aug.3, 1972, Lecture Notes in Mathematics #349, Springer-Verlag, 1973

[Do-Or] Igor DOLGACHEV and David ORTLAND, *Point Sets in Projective Spaces and Theta Functions*, Asterisque **165** (1988)

[E1] W.L. EDGE, "Klein's Encounter with the Simple Group of Order 660," Proceedings of the London Mathematical Society **24** (1972) 647-668

[E2] W.L. EDGE, "The Klein group in three dimensions," Acta Mathematica **79** (1947) 153-223

[EGA] Jean DIEUDONNÉ and Alexander GROTHENDIECK, *Éléments de Géométrie Algébrique*

[Eis1] David EISENBUD, *Commutative Algebra, with a View Toward Algebraic Geometry*, Graduate Texts in Mathematics 150, Springer-Verlag, 1995

[FGA] Alexander GROTHENDIECK, "Fondements de Geometrie Algebrique," Seminaire Bourbaki #232

[FH] Jeffrey FOX and Peter HASKELL, "Hodge decompositions and Dolbeault Complexes on Normal Surfaces," Trans. AMS **343** (1994) 765-778

[FHP] Jeffrey FOX, Peter HASKELL and William L.PARDON, "Two themes in Index Theory on Singular Variety," Proc. Symp. Pure Math. **51** (1990) pt.2 pp.103-115

[Fu1] William FULTON, *Intersection Theory*, Ergebnisse der Mathematik und ihrer Grenzgebiete, 3 Folge, Band 2, Springer-Verlag, Berlin Heidelberg New York Tokyo 1984

[Ga1] Pierre GABRIEL, "Généralités sur les groupes algébriques," Exposé VI$_A$ of SGA III, vol.1, Lecture Notes in Mathematics 151, Springer-Verlag

[Gé] P. GÉRARDIN, "Weil representations associated to finite fields," J. Alg. **42** (1976) 102-120

[Gi1] Giovanni Zeno GIAMBELLI, "Sulle varietà reppresentate coll' annullare determinanti minori contenuti in un determinante simmetrico od emisimmetrico generico di forme," Atti R. Accad. Sci. Torino **41** (1906) 102-125

[Go1] Daniel GORENSTEIN, *Finite Simple Groups: An Introduction To Their Classification*, New York: Plenum Press, 1982

[Gra] David GRANT, "Formal groups in genus 2," J. Reine Angew. Math. **411** (1990) 96-121

[Gu1] Robert Clifford GUNNING, *Lectures on Modular Forms,* Annals of Math. Studies **48**, Princeton, Princeton University Press, 1962

[Gu-Ro] Robert Clifford GUNNING and Hugo ROSSI, *Analytic Functions of Several Complex Variables,* Englewood Cliffs, NJ: Prentice-Hall, 1965

[G-N] Paul GORDAN and Max NOETHER, "Ueber die algebraischen Formen, deren Hesse'sche Determinante identisch verschwindet," Math. Ann. **10** (1876) 547-568

[Ha] Robin HARTSHORNE, *Algebraic Geometry*, Springer-Verlag, 1977, New York

[Har-Tu] Joe HARRIS and Loring W. TU, "On symmetric and skew-symmetric determinantal varieties," Topology **23** (1984) 71-84

[Has1] Peter HASKELL, "L^2-Dolbeault complexes on singular curves and surfaces," Proc. AMS **107** (1989) 517-526

[Has2] Peter HASKELL, "Index theory on curves," Trans. AMS **288** (1985) 591-604

[Has3] Peter HASKELL, "Index Theory of Geometric Fredholm Operators on Varieties with Isolated Singularities," K-theory 1 (1987) 457-466

[He1] Erich HECKE, "Die eindeutige Bestimmung der Modulfunktionen q-ter Stufe durch algebraische Eigenschaften," Math.Ann. **111** (1935) 293-301 [Mathematische Werke #31 pp.568-576]

[He2] Erich HECKE, "Ueber ein Fundamentalproblem aus der Theorie der elliptische Modulfunktionen," Abhandlungen aus dem Mathematischen Seminar des Hamburgischen Universität **6** (1928) 235-257 [Mathematische Werke art.28, p.525-547]

[Hi1] Friedrich E.P. Hirzebruch, *Topological Methods in Algebraic Geometry*, Springer Verlag

[Hi2] Friedrich E.P. Hirzebruch, "Der Satz von Riemann-Roch in faisceau theoretischer Formulierung. Einige Anwendungen und offene Fragen.", Proc. International Congress of Mathematicians 194, Vol.III, pp.457-473. Amsterdam, North Holland 1956

[Ho] Roger HOWE, "Dual reductive pairs," preprint

[Hu1] Adolf HURWITZ, "Über einige besondere homogene lineare Differentialgleichungen," Math.Ann. **26** (1886) 117-126 [Mathematische Werke, Bd.I (Funktionentheorie), Art. IX, pp.153-162]

[Hu2] Adolph HURWITZ, Über die diophantische Gleichung $x^3y + y^3z + z^3x = 0$," Math.Ann. **65** (1908) 428-430 [Math. Werke II, art.72, pp.427-429]

[Ig] Jun-Ichi IGUSA, Theta Functions, Springer Verlag 1972

[K1] Felix KLEIN, "Über die transformation siebenter Ordnung der elliptischen Funktionen," Mathematische Annalen 14 (1878) 428-471 [Ges. Math. Abh. III, art. LXXXIV, pp.90-136]

[K2] Felix KLEIN, "Über die transformation elfter Ordnung der elliptischen Funktionen," Mathematische Annalen **15** (1879) [Ges. Math. Abh. III, art.LXXXVI, pp.140-168]

[K3] Felix KLEIN, "Über gewisse Teilwerte der θ-function," Mathematische Annalen **17** (1881) 565-574 [Ges. Math. Abh. III, art. LXXXIX, pp.186-197]

[K4] Felix KLEIN, "Über die elliptischen Normalkurven der n-ten Ordnung," Abhandlungen der mathematisch-physikalischen Klassen der Sächsischen Kgl. Gesellschaft der Wissenschaften, **13**, Nr.IV (1885) [Ges. Math. Abh. III, art. XC, pp.198-254]

[K5] Felix KLEIN, "Über die Auflösung gewisser Gleichungen siebenten und achten Grades," Mathematische Annalen **15** (1879) 251-282 [Ges. Math. Abh. II, art. LVII, pp.390-425]

[K-F] Felix KLEIN and Robert FRICKE, *Vorlesungen über die Theorie der elliptischen Modulfunktionen,* Teubner, Stuttgart, 1892, p.268

[Klo] H.D.KLOOSTERMAN, "The Behaviour of General Theta Functions Under the Modular Group and the Characters of Binary Modular Congruence Groups I,II," Annals of Math. **47** (1946) 317-447

[Ko1] Shoji KOIZUMI, "The equations defining abelian varieties and modular functions," Mathematische Annalen **242** (1979)

[Ko2] Shoji KOIZUMI, "Theta relations and projective normality of abelian varieties," American Journal of Mathematics **98** (1976) 865-889

[Ku] Dan KUBERT, "Universal bounds on the torsion of elliptic curves," Proceedings of the London Mathematical Society (3) **33** (1976) 193-237

[Mas] Heinrich MASCHKE, "Aufstellung des vollen Formensystem einer quaternären Gruppe vom 51840 Substitutionen," Mathematische Annalen **33** (1889) 317-344

[Maz] Barry MAZUR, "Modular Curves and the Eisenstein Ideal," Publications Math. IHES **47** (1977) 33-186

[Mi] John MILNOR, *Morse Theory,* Annals of Mathematics Studies 51, Princeton University Press, 1963

[Mu1] David MUMFORD, "On the equations defining abelian varieties," I, Inv. Math. **1** (1966) 287-354; II, Inv. Math. **3** (1967) 73-135; III, Inv. Math. **3** (1967) 215-244

[Mu2] David MUMFORD, "Varieties defined by quadratic equations," Questions on Algebraic Varieties, CIME, 1970, pp. 29-100

[Mu3] David MUMFORD, "The structure of the moduli space of curves and abelian varieties," Actes Congre. Intern. Math. 1970, Tome I, 457-465

[Mu4] David MUMFORD, Appendix to *Geometric Invariant Theory,* 2nd ed."

[Mu5] David MUMFORD, *Abelian Varieties,* Oxford University Press, London and Tata Institute of Fundamental Research, Bombay, 1970

[N-R 1] M.S.NARASIMHAN and Sundararaman RAMANAN, "2θ-linear systems on abelian varieties," in Vector Bundles on Algebraic Varieties, Bombay Colloquium 1984, published for the Tata Institute of Fundamental Research, Bombay by Oxford University Press, Bombay, Delhi, Calcutta, Madras 1987

[N-R 2] M.S.NARASIMHAN and Sundararaman RAMANAN, "Vector bundles on Curves," Algebraic Geometry, Papers Presented at the Bombay Colloquium, 1968, published by Tata Institute of Fundamental Research and Oxford University Press, 1969

[Pa-St] William L. PARDON and Mark A. STERN, "$L^2 - \bar{\partial}$-cohomology of complex projective varieties," J.AMS **4** (1991) 603-621

[Po-Gr1] Sorin POPESCU and Mark GROSS, "Equations of $(1, d)$-polarized abelian surfaces," preprint alg-geom/9609001, available from eprints.math.duke.edu.

[Po-Gr2] Sorin POPESCU and Mark GROSS, "Calabi-Yau 3-folds and moduli of abelian surfaces", in preparation.

[Sa1] Leslie SAPER, "L_2-cohomology of Kähler varieties with isolated singularities," J. Diff. Geom. **36** (1992) 89-161

[Sch1] R.L.E.SCHWARZENBERGER, Appendix One of [Hi1]

[Sek] Tsutomu SEKIGUCHI, "On the cubics defining abelian varieties," Journal of the Mathematical Society of Japan **30** (1978) 703-721

[SGA1] Alexander GROTHENDIECK, *SGA 1: Revêtements Etales et Groupe Fondamental*, Lecture Notes in Mathematics 224, Springer-Verlag, 1971

[Shi1] Hideo SHIMIZU, "On zeta functions of quaternion algebras," Ann. Math. **81** (1965) 166-193

[Shu1] Goro SHIMURA, "Construction of class fields and zeta functions of algebraic curves," Annals of Mathematics **85** (1967) 58-159

[Shu2] Goro SHIMURA, *The arithmetic theory of automorphic forms,* Iwanami Shoten and Princeton University Press

[St1] Mark A. STERN, "L_2-cohomology and index theory of noncompact manifolds," Proc. Symp. in Pure Math. **54** (1993) pt.2 pp.559-575

[St2] Mark A. STERN, "Index theory for certain complete Kähler manifolds," J. Diff. Geom. **37** (1993) 467-503

[T1] Shun'ichi TANAKA, "On Irreducible Unitary Representations of Some Special Linear Groups of the Second Order I," Osaka J. Math. **3** (1966) 217-227

[T2] Shun'ichi TANAKA, "On Irreducible Unitary Representations of Some Special Linear Groups of the Second Order II," Osaka J. Math. **3** (1966) 229-242

[T3] Shun'ichi TANAKA, "Construction and Classification of Irreducible Representations of Special Linear Group of the Second Order Over a Finite Field," Osaka J. Math.**4** (1967) 65-84

[T4] Shun'ichi TANAKA, "Irreducible Representation of the Binary Modular Congruence Groups Mod p^λ," J. Math. Kyoto University **7-2** (1967)123-132

[Ta-Wi] Richard TAYLOR and Andrew WILES, "Ring thoeretic properties of certain Hecke algebras," Ann.Math. **141** (1995) 553-572

[VdG1] Gerard VAN DER GEER, "On the geometry of a Siegel modular threefold," Math. Ann. **260** (1982) 317-350

[VdG2] G.VAN DER GEER, "Note on some abelian schemes of level three," Math.Ann. **278** (1987) 401-408

[Vé 1] Jacques VÉLU, "Courbes elliptiques munies d'un sousgroupe $\mathbf{Z}/n\mathbf{Z} \times \mu_n$," Thèse, Université Paris Sud, 1978

[Vé 2] Jacques VÉLU, "Courbes modulaires et courbe de Fermat," Sem. de la Théorie des Nombres (Univ. Bordeaux, Talence) Exp. 16, CNRS, Talence 1976

[Vé3] Jacques VÉLU, Table 3 of Modular Functions of One Variable IV, Springer Lecture Notes in Mathematics #476, Springer-Verlag 1975

[W1] André WEIL, "Sur certains groupes d'operateurs unitaires," Acta Mathematica **111** (1964) 143-211, Oeuvres Scientifiques, Vol.III, art. [1964b], 1-69

[W2] André WEIL, *Foundations of Algebraic Geometry,* American Mathematical Society Colloquium Publications XXIX, Rev. and enlarged edition, 1962, American Mathematical Society, Providence, RI

[W3] André WEIL, *Variétés Kählériennes,* Actualités scientifiques et industrielles 1267 Publication de l'Institut de Mathématique de l'Université de Nancago VI Hermann, Paris 1971

[W4] André WEIL, "Sur la formule de Siegel dans la théorie des groupes classiques," Acta Math. **113** (1965) 1-87 [Oeuvres Scientifiques III, art. [1965]]

[W1] André WEIL, "On Picard varieties," Am.J.Math. **74** (1952) 865-894 [Oeuvres Scientifiques, art. [1952a], Vol.II pp.73-102]

[Wi1] Andrew WILES, "Modular elliptic curves and Fermat's last theorem," Ann. Math. **141** (1995) 443-551

Index of Notation

For the convenience of the reader, we have provided below an index of the notation used in this book:

Subject Index

Springer
and the
environment

At Springer we firmly believe that an
international science publisher has a
special obligation to the environment,
and our corporate policies consistently
reflect this conviction.
We also expect our business partners –
paper mills, printers, packaging
manufacturers, etc. – to commit
themselves to using materials and
production processes that do not harm
the environment. The paper in this
book is made from low- or no-chlorine
pulp and is acid free, in conformance
with international standards for paper
permanency.

 Springer

Lecture Notes in Mathematics

For information about Vols. 1–1459
please contact your bookseller or Springer-Verlag

Vol. 1500: J.-P. Serre, Lie Algebras and Lie Groups. VII, 168 pages. 1992.

Vol. 1501: A. De Masi, E. Presutti, Mathematical Methods for Hydrodynamic Limits. IX, 196 pages. 1991.

Vol. 1502: C. Simpson, Asymptotic Behavior of Monodromy. V, 139 pages. 1991.

Vol. 1503: S. Shokranian, The Selberg-Arthur Trace Formula (Lectures by J. Arthur). VII, 97 pages. 1991.

Vol. 1504: J. Cheeger, M. Gromov, C. Okonek, P. Pansu, Geometric Topology: Recent Developments. Editors: P. de Bartolomeis, F. Tricerri. VII, 197 pages. 1991.

Vol. 1505: K. Kajitani, T. Nishitani, The Hyperbolic Cauchy Problem. VII, 168 pages. 1991.

Vol. 1506: A. Buium, Differential Algebraic Groups of Finite Dimension. XV, 145 pages. 1992.

Vol. 1507: K. Hulek, T. Peternell, M. Schneider, F.-O. Schreyer (Eds.), Complex Algebraic Varieties. Proceedings, 1990. VII, 179 pages. 1992.

Vol. 1508: M. Vuorinen (Ed.), Quasiconformal Space Mappings. A Collection of Surveys 1960-1990. IX, 148 pages. 1992.

Vol. 1509: J. Aguadé, M. Castellet, F. R. Cohen (Eds.), Algebraic Topology - Homotopy and Group Cohomology. Proceedings, 1990. X, 330 pages. 1992.

Vol. 1510: P. P. Kulish (Ed.), Quantum Groups. Proceedings, 1990. XII, 398 pages. 1992.

Vol. 1511: B. S. Yadav, D. Singh (Eds.), Functional Analysis and Operator Theory. Proceedings, 1990. VIII, 223 pages. 1992.

Vol. 1512: L. M. Adleman, M.-D. A. Huang, Primality Testing and Abelian Varieties Over Finite Fields. VII, 142 pages. 1992.

Vol. 1513: L. S. Block, W. A. Coppel, Dynamics in One Dimension. VIII, 249 pages. 1992.

Vol. 1514: U. Krengel, K. Richter, V. Warstat (Eds.), Ergodic Theory and Related Topics III, Proceedings, 1990. VIII, 236 pages. 1992.

Vol. 1515: E. Ballico, F. Catanese, C. Ciliberto (Eds.), Classification of Irregular Varieties. Proceedings, 1990. VII, 149 pages. 1992.

Vol. 1516: R. A. Lorentz, Multivariate Birkhoff Interpolation. IX, 192 pages. 1992.

Vol. 1517: K. Keimel, W. Roth, Ordered Cones and Approximation. VI, 134 pages. 1992.

Vol. 1518: H. Stichtenoth, M. A. Tsfasman (Eds.), Coding Theory and Algebraic Geometry. Proceedings, 1991. VIII, 223 pages. 1992.

Vol. 1519: M. W. Short, The Primitive Soluble Permutation Groups of Degree less than 256. IX, 145 pages. 1992.

Vol. 1520: Yu. G. Borisovich, Yu. E. Gliklikh (Eds.), Global Analysis – Studies and Applications V. VII, 284 pages. 1992.

Vol. 1521: S. Busenberg, B. Forte, H. K. Kuiken, Mathematical Modelling of Industrial Process. Bari, 1990. Editors: V. Capasso, A. Fasano. VII, 162 pages. 1992.

Vol. 1522: J.-M. Delort, F. B. I. Transformation. VII, 101 pages. 1992.

Vol. 1523: W. Xue, Rings with Morita Duality. X, 168 pages. 1992.

Vol. 1524: M. Coste, L. Mahé, M.-F. Roy (Eds.), Real Algebraic Geometry. Proceedings, 1991. VIII, 418 pages. 1992.

Vol. 1525: C. Casacuberta, M. Castellet (Eds.), Mathematical Research Today and Tomorrow. VII, 112 pages. 1992.

Vol. 1526: J. Azéma, P. A. Meyer, M. Yor (Eds.), Séminaire de Probabilités XXVI. X, 633 pages. 1992.

Vol. 1527: M. I. Freidlin, J.-F. Le Gall, Ecole d'Eté de Probabilités de Saint-Flour XX – 1990. Editor: P. L. Hennequin. VIII, 244 pages. 1992.

Vol. 1528: G. Isac, Complementarity Problems. VI, 297 pages. 1992.

Vol. 1529: J. van Neerven, The Adjoint of a Semigroup of Linear Operators. X, 195 pages. 1992.

Vol. 1530: J. G. Heywood, K. Masuda, R. Rautmann, S. A. Solonnikov (Eds.), The Navier-Stokes Equations II – Theory and Numerical Methods. IX, 322 pages. 1992.

Vol. 1531: M. Stoer, Design of Survivable Networks. IV, 206 pages. 1992.

Vol. 1532: J. F. Colombeau, Multiplication of Distributions. X, 184 pages. 1992.

Vol. 1533: P. Jipsen, H. Rose, Varieties of Lattices. X, 162 pages. 1992.

Vol. 1534: C. Greither, Cyclic Galois Extensions of Commutative Rings. X, 145 pages. 1992.

Vol. 1535: A. B. Evans, Orthomorphism Graphs of Groups. VIII, 114 pages. 1992.

Vol. 1536: M. K. Kwong, A. Zettl, Norm Inequalities for Derivatives and Differences. VII, 150 pages. 1992.

Vol. 1537: P. Fitzpatrick, M. Martelli, J. Mawhin, R. Nussbaum, Topological Methods for Ordinary Differential Equations. Montecatini Terme, 1991. Editors: M. Furi, P. Zecca. VII, 218 pages. 1993.

Vol. 1538: P.-A. Meyer, Quantum Probability for Probabilists. X, 287 pages. 1993.

Vol. 1539: M. Coornaert, A. Papadopoulos, Symbolic Dynamics and Hyperbolic Groups. VIII, 138 pages. 1993.

Vol. 1540: H. Komatsu (Ed.), Functional Analysis and Related Topics, 1991. Proceedings. XXI, 413 pages. 1993.

Vol. 1541: D. A. Dawson, B. Maisonneuve, J. Spencer, Ecole d´Eté de Probabilités de Saint-Flour XXI - 1991. Editor: P. L. Hennequin. VIII, 356 pages. 1993.

Vol. 1542: J.Fröhlich, Th.Kerler, Quantum Groups, Quantum Categories and Quantum Field Theory. VII, 431 pages. 1993.

Vol. 1543: A. L. Dontchev, T. Zolezzi, Well-Posed Optimization Problems. XII, 421 pages. 1993.

Vol. 1544: M.Schürmann, White Noise on Bialgebras. VII, 146 pages. 1993.

Vol. 1545: J. Morgan, K. O'Grady, Differential Topology of Complex Surfaces. VIII, 224 pages. 1993.

Vol. 1546: V. V. Kalashnikov, V. M. Zolotarev (Eds.), Stability Problems for Stochastic Models. Proceedings, 1991. VIII, 229 pages. 1993.

Vol. 1547: P. Harmand, D. Werner, W. Werner, M-ideals in Banach Spaces and Banach Algebras. VIII, 387 pages. 1993.

Vol. 1548: T. Urabe, Dynkin Graphs and Quadrilateral Singularities. VI, 233 pages. 1993.

Vol. 1549: G. Vainikko, Multidimensional Weakly Singular Integral Equations. XI, 159 pages. 1993.

Vol. 1550: A. A. Gonchar, E. B. Saff (Eds.), Methods of Approximation Theory in Complex Analysis and Mathematical Physics IV, 222 pages, 1993.

Vol. 1551: L. Arkeryd, P. L. Lions, P.A. Markowich, S.R. S. Varadhan. Nonequilibrium Problems in Many-Particle Systems. Montecatini, 1992. Editors: C. Cercignani, M. Pulvirenti. VII, 158 pages 1993.

Vol. 1552: J. Hilgert, K.-H. Neeb, Lie Semigroups and their Applications. XII, 315 pages. 1993.

Vol. 1553: J.-L- Colliot-Thélène, J. Kato, P. Vojta. Arithmetic Algebraic Geometry. Trento, 1991. Editor: E. Ballico. VII, 223 pages. 1993.

Vol. 1554: A. K. Lenstra, H. W. Lenstra, Jr. (Eds.), The Development of the Number Field Sieve. VIII, 131 pages. 1993.

Vol. 1555: O. Liess, Conical Refraction and Higher Microlocalization. X, 389 pages. 1993.

Vol. 1556: S. B. Kuksin, Nearly Integrable Infinite-Dimensional Hamiltonian Systems. XXVII, 101 pages. 1993.

Vol. 1557: J. Azéma, P. A. Meyer, M. Yor (Eds.), Séminaire de Probabilités XXVII. VI, 327 pages. 1993.

Vol. 1558: T. J. Bridges, J. E. Furter, Singularity Theory and Equivariant Symplectic Maps. VI, 226 pages. 1993.

Vol. 1559: V. G. Sprindžuk, Classical Diophantine Equations. XII, 228 pages. 1993.

Vol. 1560: T. Bartsch, Topological Methods for Variational Problems with Symmetries. X, 152 pages. 1993.

Vol. 1561: I. S. Molchanov, Limit Theorems for Unions of Random Closed Sets. X, 157 pages. 1993.

Vol. 1562: G. Harder, Eisensteinkohomologie und die Konstruktion gemischter Motive. XX, 184 pages. 1993.

Vol. 1563: E. Fabes, M. Fukushima, L. Gross, C. Kenig, M. Röckner, D. W. Stroock, Dirichlet Forms. Varenna, 1992. Editors: G. Dell'Antonio, U. Mosco. VII, 245 pages. 1993.

Vol. 1564: J. Jorgenson, S. Lang, Basic Analysis of Regularized Series and Products. IX, 122 pages. 1993.

Vol. 1565: L. Boutet de Monvel, C. De Concini, C. Procesi, P. Schapira, M. Vergne. D-modules, Representation Theory, and Quantum Groups. Venezia, 1992. Editors: G. Zampieri, A. D'Agnolo. VII, 217 pages. 1993.

Vol. 1566: B. Edixhoven, J.-H. Evertse (Eds.), Diophantine Approximation and Abelian Varieties. XIII, 127 pages. 1993.

Vol. 1567: R. L. Dobrushin, S. Kusuoka, Statistical Mechanics and Fractals. VII, 98 pages. 1993.

Vol. 1568: F. Weisz, Martingale Hardy Spaces and their Application in Fourier Analysis. VIII, 217 pages. 1994.

Vol. 1569: V. Totik, Weighted Approximation with Varying Weight. VI, 117 pages. 1994.

Vol. 1570: R. deLaubenfels, Existence Families, Functional Calculi and Evolution Equations. XV, 234 pages. 1994.

Vol. 1571: S. Yu. Pilyugin, The Space of Dynamical Systems with the C^0-Topology. X, 188 pages. 1994.

Vol. 1572: L. Göttsche, Hilbert Schemes of Zero-Dimensional Subschemes of Smooth Varieties. IX, 196 pages. 1994.

Vol. 1573: V. P. Havin, N. K. Nikolski (Eds.), Linear and Complex Analysis – Problem Book 3 – Part I. XXII, 489 pages. 1994.

Vol. 1574: V. P. Havin, N. K. Nikolski (Eds.), Linear and Complex Analysis – Problem Book 3 – Part II. XXII, 507 pages. 1994.

Vol. 1575: M. Mitrea, Clifford Wavelets, Singular Integrals, and Hardy Spaces. XI, 116 pages. 1994.

Vol. 1576: K. Kitahara, Spaces of Approximating Functions with Haar-Like Conditions. X, 110 pages. 1994.

Vol. 1577: N. Obata, White Noise Calculus and Fock Space. X, 183 pages. 1994.

Vol. 1578: J. Bernstein, V. Lunts, Equivariant Sheaves and Functors. V, 139 pages. 1994.

Vol. 1579: N. Kazamaki, Continuous Exponential Martingales and BMO. VII, 91 pages. 1994.

Vol. 1580: M. Milman, Extrapolation and Optimal Decompositions with Applications to Analysis. XI, 161 pages. 1994.

Vol. 1581: D. Bakry, R. D. Gill, S. A. Molchanov, Lectures on Probability Theory. Editor: P. Bernard. VIII, 420 pages. 1994.

Vol. 1582: W. Balser, From Divergent Power Series to Analytic Functions. X, 108 pages. 1994.

Vol. 1583: J. Azéma, P. A. Meyer, M. Yor (Eds.), Séminaire de Probabilités XXVIII. VI, 334 pages. 1994.

Vol. 1584: M. Brokate, N. Kenmochi, I. Müller, J. F. Rodriguez, C. Verdi, Phase Transitions and Hysteresis. Montecatini Terme, 1993. Editor: A. Visintin. VII. 291 pages. 1994.

Vol. 1585: G. Frey (Ed.), On Artin's Conjecture for Odd 2-dimensional Representations. VIII, 148 pages. 1994.

Vol. 1586: R. Nillsen, Difference Spaces and Invariant Linear Forms. XII, 186 pages. 1994.

Vol. 1587: N. Xi, Representations of Affine Hecke Algebras. VIII, 137 pages. 1994.

Vol. 1588: C. Scheiderer, Real and Étale Cohomology. XXIV, 273 pages. 1994.

Vol. 1589: J. Bellissard, M. Degli Esposti, G. Forni, S. Graffi, S. Isola, J. N. Mather, Transition to Chaos in Classical and Quantum Mechanics. Montecatini Terme, 1991. Editor: S. Graffi. VII, 192 pages. 1994.

Vol. 1590: P. M. Soardi, Potential Theory on Infinite Networks. VIII, 187 pages. 1994.

Vol. 1591: M. Abate, G. Patrizio, Finsler Metrics – A Global Approach. IX, 180 pages. 1994.

Vol. 1592: K. W. Breitung, Asymptotic Approximations for Probability Integrals. IX, 146 pages. 1994.

Vol. 1593: J. Jorgenson & S. Lang, D. Goldfeld, Explicit Formulas for Regularized Products and Series. VIII, 154 pages. 1994.

Vol. 1594: M. Green, J. Murre, C. Voisin, Algebraic Cycles and Hodge Theory. Torino, 1993. Editors: A. Albano, F. Bardelli. VII, 275 pages. 1994.

Vol. 1595: R.D.M. Accola, Topics in the Theory of Riemann Surfaces. IX, 105 pages. 1994.

Vol. 1596: L. Heindorf, L. B. Shapiro, Nearly Projective Boolean Algebras. X, 202 pages. 1994.

Vol. 1597: B. Herzog, Kodaira-Spencer Maps in Local Algebra. XVII, 176 pages. 1994.

Vol. 1598: J. Berndt, F. Tricerri, L. Vanhecke, Generalized

Heisenberg Groups and Damek-Ricci Harmonic Spaces. VIII, 125 pages. 1995.

Vol. 1599: K. Johannson, Topology and Combinatorics of 3-Manifolds. XVIII, 446 pages. 1995.

Vol. 1600: W. Narkiewicz, Polynomial Mappings. VII, 130 pages. 1995.

Vol. 1601: A. Pott, Finite Geometry and Character Theory. VII, 181 pages. 1995.

Vol. 1602: J. Winkelmann, The Classification of Three-dimensional Homogeneous Complex Manifolds. XI, 230 pages. 1995.

Vol. 1603: V. Ene, Real Functions – Current Topics. XIII, 310 pages. 1995.

Vol. 1604: A. Huber, Mixed Motives and their Realization in Derived Categories. XV, 207 pages. 1995.

Vol. 1605: L. B. Wahlbin, Superconvergence in Galerkin Finite Element Methods. XI, 166 pages. 1995.

Vol. 1606: P.-D. Liu, M. Qian, Smooth Ergodic Theory of Random Dynamical Systems. XI, 221 pages. 1995.

Vol. 1607: G. Schwarz, Hodge Decomposition – A Method for Solving Boundary Value Problems. VII, 155 pages. 1995.

Vol. 1608: P. Biane, R. Durrett, Lectures on Probability Theory. Editor: P. Bernard. VII, 210 pages. 1995.

Vol. 1609: L. Arnold, C. Jones, K. Mischaikow, G. Raugel, Dynamical Systems. Montecatini Terme, 1994. Editor: R. Johnson. VIII, 329 pages. 1995.

Vol. 1610: A. S. Üstünel, An Introduction to Analysis on Wiener Space. X, 95 pages. 1995.

Vol. 1611: N. Knarr, Translation Planes. VI, 112 pages. 1995.

Vol. 1612: W. Kühnel, Tight Polyhedral Submanifolds and Tight Triangulations. VII, 122 pages. 1995.

Vol. 1613: J. Azéma, M. Emery, P. A. Meyer, M. Yor (Eds.), Séminaire de Probabilités XXIX. VI, 326 pages. 1995.

Vol. 1614: A. Koshelev, Regularity Problem for Quasilinear Elliptic and Parabolic Systems. XXI, 255 pages. 1995.

Vol. 1615: D. B. Massey, Lê Cycles and Hypersurface Singularities. XI, 131 pages. 1995.

Vol. 1616: I. Moerdijk, Classifying Spaces and Classifying Topoi. VII, 94 pages. 1995.

Vol. 1617: V. Yurinsky, Sums and Gaussian Vectors. XI, 305 pages. 1995.

Vol. 1618: G. Pisier, Similarity Problems and Completely Bounded Maps. VII, 156 pages. 1996.

Vol. 1619: E. Landvogt, A Compactification of the Bruhat-Tits Building. VII, 152 pages. 1996.

Vol. 1620: R. Donagi, B. Dubrovin, E. Frenkel, E. Previato, Integrable Systems and Quantum Groups. Montecatini Terme, 1993. Editors: M. Francaviglia, S. Greco. VIII, 488 pages. 1996.

Vol. 1621: H. Bass, M. V. Otero-Espinar, D. N. Rockmore, C. P. L. Tresser, Cyclic Renormalization and Auto-morphism Groups of Rooted Trees. XXI, 136 pages. 1996.

Vol. 1622: E. D. Farjoun, Cellular Spaces, Null Spaces and Homotopy Localization. XIV, 199 pages. 1996.

Vol. 1623: H.P. Yap, Total Colourings of Graphs. VIII, 131 pages. 1996.

Vol. 1624: V. Brînzănescu, Holomorphic Vector Bundles over Compact Complex Surfaces. X, 170 pages. 1996.

Vol. 1625: S. Lang, Topics in Cohomology of Groups. VII, 226 pages. 1996.

Vol. 1626: J. Azéma, M. Emery, M. Yor (Eds.), Séminaire de Probabilités XXX. VIII, 382 pages. 1996.

Vol. 1627: C. Graham, Th. G. Kurtz, S. Méléard, Ph. E. Protter, M. Pulvirenti, D. Talay, Probabilistic Models for Nonlinear Partial Differential Equations. Montecatini Terme, 1995. Editors: D. Talay, L. Tubaro. X, 301 pages. 1996.

Vol. 1628: P.-H. Zieschang, An Algebraic Approach to Association Schemes. XII, 189 pages. 1996.

Vol. 1629: J. D. Moore, Lectures on Seiberg-Witten Invariants. VII, 105 pages. 1996.

Vol. 1630: D. Neuenschwander, Probabilities on the Heisenberg Group: Limit Theorems and Brownian Motion. VIII, 139 pages. 1996.

Vol. 1631: K. Nishioka, Mahler Functions and Transcendence. VIII, 185 pages. 1996.

Vol. 1632: A. Kushkuley, Z. Balanov, Geometric Methods in Degree Theory for Equivariant Maps. VII, 136 pages. 1996.

Vol. 1633: H. Aikawa, M. Essén, Potential Theory – Selected Topics. IX, 200 pages. 1996.

Vol. 1634: J. Xu, Flat Covers of Modules. IX, 161 pages. 1996.

Vol. 1635: E. Hebey, Sobolev Spaces on Riemannian Manifolds. X, 116 pages. 1996.

Vol. 1636: M. A. Marshall, Spaces of Orderings and Abstract Real Spectra. VI, 190 pages. 1996.

Vol. 1637: B. Hunt, The Geometry of some special Arithmetic Quotients. XIII, 332 pages. 1996.

Vol. 1638: P. Vanhaecke, Integrable Systems in the realm of Algebraic Geometry. VIII, 218 pages. 1996.

Vol. 1639: K. Dekimpe, Almost-Bieberbach Groups: Affine and Polynomial Structures. X, 259 pages. 1996.

Vol. 1640: G. Boillat, C. M. Dafermos, P. D. Lax, T. P. Liu, Recent Mathematical Methods in Nonlinear Wave Propagation. Montecatini Terme, 1994. Editor: T. Ruggeri. VII, 142 pages. 1996.

Vol. 1641: P. Abramenko, Twin Buildings and Applications to S-Arithmetic Groups. IX, 123 pages. 1996.

Vol. 1642: M. Puschnigg, Asymptotic Cyclic Cohomology. XXII, 138 pages. 1996.

Vol. 1643: J. Richter-Gebert, Realization Spaces of Polytopes. XI, 187 pages. 1996.

Vol. 1644: A. Adler, S. Ramanan, Moduli of Abelian Varieties. VI, 196 pages. 1996.

Vol. 1645: H. W. Broer, G. B. Huitema, M. B. Sevryuk, Quasi-Periodic Motions in Families of Dynamical Systems. XI, 195 pages. 1996.

Vol. 1646: J.-P. Demailly, T. Peternell, G. Tian, A. N. Tyurin, Transcendental Methods in Algebraic Geometry. Cetraro, 1994. Editors: F. Catanese, C. Ciliberto. VII, 257 pages. 1996.

Vol. 1647: D. Dias, P. Le Barz, Configuration Spaces over Hilbert Schemes and Applications. VII, 143 pages. 1996.

Vol. 1648: R. Dobrushin, P. Groeneboom, M. Ledoux, Lectures on Probability Theory and Statistics. Editor: P. Bernard. VIII, 300 pages. 1996.

Druck: Weihert-Druck GmbH, Darmstadt
Bindearbeiten: Theo Gansert Buchbinderei GmbH, Weinheim